Basic Sensors in iOS

Basic Sensors in iOS

Alasdair Allan

Beijing · Cambridge · Farnham · Köln · Sebastopol · Tokyo

Basic Sensors in iOS
by Alasdair Allan

Copyright © 2011 Alasdair Allan. All rights reserved.
Printed in the United States of America.

Published by O'Reilly Media, Inc., 1005 Gravenstein Highway North, Sebastopol, CA 95472.

O'Reilly books may be purchased for educational, business, or sales promotional use. Online editions are also available for most titles (*http://my.safaribooksonline.com*). For more information, contact our corporate/institutional sales department: (800) 998-9938 or *corporate@oreilly.com*.

Editor: Brian Jepson
Proofreader: O'Reilly Production Services

Cover Designer: Karen Montgomery
Interior Designer: David Futato
Illustrator: Robert Romano

Printing History:
> July 2011:   First Edition.

ISBN: 978-1-449-30846-9

[LSI]

1311179613

Table of Contents

Preface vii

1. **The Hardware** 1
   Available Sensor Hardware   1
   Differences Between iPhone and iPad   2
      Device Orientation and the iPad   4
   Detecting Hardware Differences   4
      Camera Availability   5
      Audio Input Availability   5
      GPS Availability   6
      Magnetometer Availability   6
   Setting Required Hardware Capabilities   6
      Persistent WiFi   7
      Background Modes   7

2. **Using the Camera** 9
   The Hardware   9
   Capturing Stills and Video   10
   Video Thumbnails   18
      Video Thumbnails Using the UIImagePicker   18
      Video Thumbnails Using AVFoundation   19
   Saving Media to the Photo Album   20
   Video Customization   23

3. **Using Audio** 25
   The Hardware   25
   Media Playback   26
   Recording and Playing Audio   31
      Recording Audio   32
      Playing Audio   35

4. Using the Accelerometer **37**

About the Accelerometer 37

Writing an Accelerometer Application 38

Determining Device Orientation 43

Determining Device Orientation Directly Using the Accelerometer 46

Obtaining Notifications when Device Orientation Changes 48

Which Way Is Up? 49

Convenience Methods for Orientation 52

Detecting Shaking 53

5. Using the Magnetometer **57**

About the Magnetometer 57

Writing a Magnetometer Application 59

Determining the Heading in Landscape Mode 62

Measuring a Magnetic Field 68

6. Using Core Motion **71**

Core Motion 71

Pulling Motion Data 72

Pushing Motion Data 73

Accessing the Gyroscope 75

Measuring Device Motion 79

Comparing Device Motion with the Accelerometer 83

7. Going Further **87**

The iPhone SDK 87

Geolocation and Maps 87

Third-Party SDKs 87

Speech Recognition 88

Computer Vision 88

Augmented Reality 88

External Accessories 88

Preface

Over the last few years the new generation of smart phones, such as Apple's iPhone, has finally started to live up to their name and have become the primary interface device for geographically tagged data. However not only do these devices know where they are, they can tell you how they're being held, they are sufficiently powerful to overlay data layers on the camera view, and record and interpret audio data, and they can do all this in real time. These are not just smart phones, these are computers that just happen to be able to make phone calls.

This book should provide a solid introduction to using the hardware features in the iPhone, iPod touch, and iPad.

Who Should Read This Book?

This book provides an introduction to the hot topic of location-enabled sensors on the iPhone. If you are a programmer who has had some experience with the iPhone before, this book will help you push your knowledge further. If you are an experienced Mac programmer, already familiar with Objective-C as a language, this book will give you an introduction to the hardware specific parts of iPhone programming.

What You Should Already Know?

The book assumes some previous experience with the Objective-C language. Additionally some familiarity with the iPhone platform would be helpful. If you're new to the iPhone platform you may be interested in *Learning iPhone Programming* (*http://oreilly.com/catalog/9780596806439*), also by Alasdair Allan (O'Reilly).

What Will You Learn?

This book will guide you through guide you through developing applications for the iPhone platform that make use of the onboard sensors: the three-axis accelerometer, the magnetometer (digital compass), the gyroscope, the camera and the global positioning system

What's In This Book?

Chapter 1, *The Hardware*
> This chapter summarizes the available sensors on the iPhone and iPad platforms and how they have, or could be, used in applications. It talks about the differences between the hardware platforms.

Chapter 2, *Using the Camera*
> Walkthrough of how to use the iPhone's camera for still and video. How to create video thumbnails and customise video.

Chapter 3, *Using Audio*
> Walkthrough of how to playback iPod media, as well as how to play and record audio on your device.

Chapter 4, *Using the Accelerometer*
> Walkthrough of how to use the accelerometer, discussion of what is implied with respect to the orientation of the device by the raw readings.

Chapter 5, *Using the Magnetometer*
> Walkthrough of how to use the magnetometer, discussion of combining the magnetometer and accelerometer to get the yaw, pitch and roll of the device.

Chapter 6, *Using Core Motion*
> This paragraph discusses the new Core Motion framework; this new framework allows your application to receive motion data from both the accelerometer and (on the latest generation of devices) the gyroscope.

Chapter 7, *Going Further*
> Provides a collection of pointers to more advanced material on the topics we covered in the book, and material covering some of those topics that we didn't manage to talk about in this book.

Conventions Used in This Book

The following typographical conventions are used in this book:

Italic
> Indicates new terms, URLs, email addresses, filenames, and file extensions.

`Constant width`
> Used for program listings, as well as within paragraphs to refer to program elements such as variable or function names, databases, data types, environment variables, statements, and keywords.

Constant width bold

Shows commands or other text that should be typed literally by the user.

Constant width italic

Shows text that should be replaced with user-supplied values or by values determined by context.

 This icon signifies a tip, suggestion, or general note.

 This icon signifies a warning or caution.

Using Code Examples

This book is here to help you get your job done. In general, you may use the code in this book in your programs and documentation. You do not need to contact us for permission unless you're reproducing a significant portion of the code. For example, writing a program that uses several chunks of code from this book does not require permission. Selling or distributing a CD-ROM of examples from O'Reilly books does require permission. Answering a question by citing this book and quoting example code does not require permission. Incorporating a significant amount of example code from this book into your product's documentation does require permission.

We appreciate, but do not require, attribution. An attribution usually includes the title, author, publisher, and ISBN. For example: "*Basic Sensors in iOS*, by Alasdair Allan. Copyright 2011 O'Reilly Media, Inc., 978-1-4493-0846-9."

If you feel your use of code examples falls outside fair use or the permission given here, feel free to contact us at *permissions@oreilly.com*.

 A lot of the examples won't work completely in the simulator, so you'll need to deploy them to your device to test the code.

Safari® Books Online

 Safari Books Online is an on-demand digital library that lets you easily search over 7,500 technology and creative reference books and videos to find the answers you need quickly.

With a subscription, you can read any page and watch any video from our library online. Read books on your cell phone and mobile devices. Access new titles before they are available for print, and get exclusive access to manuscripts in development and post feedback for the authors. Copy and paste code samples, organize your favorites, download chapters, bookmark key sections, create notes, print out pages, and benefit from tons of other time-saving features.

O'Reilly Media has uploaded this book to the Safari Books Online service. To have full digital access to this book and others on similar topics from O'Reilly and other publishers, sign up for free at *http://my.safaribooksonline.com*.

How to Contact Us

Please address comments and questions concerning this book to the publisher:

O'Reilly Media, Inc.
1005 Gravenstein Highway North
Sebastopol, CA 95472
800-998-9938 (in the United States or Canada)
707-829-0515 (international or local)
707-829-0104 (fax)

We have a web page for this book, where we list errata, examples, and any additional information. You can access this page at:

http://www.oreilly.com/catalog/9781449308469

Supplementary materials are also available at:

http://www.programmingiphonesensors.com

To comment or ask technical questions about this book, send email to:

bookquestions@oreilly.com

For more information about our books, courses, conferences, and news, see our website at *http://www.oreilly.com*.

Find us on Facebook: *http://facebook.com/oreilly*

Follow us on Twitter: *http://twitter.com/oreillymedia*

Watch us on YouTube: *http://www.youtube.com/oreillymedia*

Acknowledgments

Everyone has one book in them. This is my second, or depending how you want to look at it, my Platform 9¾, since this book, along with the other three forthcoming short books on iOS and sensor technology, will form the bulk of *Programming iOS4 Sensors*, which would probably be classed by most as my second real book for O'Reilly.

At which point, I'd like to thank my editor at O'Reilly, Brian Jepson, for holding my hand just one more time. As ever his hard work made my hard work much better than it otherwise would have been. I also very much want to thank my wife Gemma Hobson for her continued support and encouragement. Those small, and sometimes larger, sacrifices an author's spouse routinely has to make don't get any less inconvenient the second time around. I'm not sure why she let me write another, perhaps because I claimed to enjoy writing the first one so much. Thank you Gemma. Finally to my son Alex, still too young to read what his daddy has written, hopefully this volume will keep you in books to chew on just a little longer.

CHAPTER 1

The Hardware

The arrival of the iPhone changed the whole direction of software development for mobile platforms, and has had a profound impact on the hardware design of the smart phones that have followed it. The arrival of the iPad has turned what was a single class of device into a platform.

Available Sensor Hardware

While the iPhone is almost unique amongst mobile platforms in guaranteeing that your application will run on all of the current devices (see Figure 1-1), however there is an increasing amount of variation in available hardware between the various models, as shown in Table 1-1.

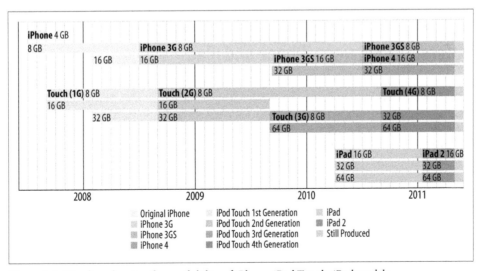

Figure 1-1. Timeline showing the availability of iPhone, iPod Touch, iPad models

Table 1-1. *Hardware support in various iPhone, iPod touch, and iPad*

Hardware Feature	iPhone				iPod touch				iPad		iPad 2	
	Original	3G	3GS	4	1st Gen	2nd Gen	3rd Gen	4th Gen	WiFi	3G	WiFi	3G
Cellular	☑	☑	☑	☑	☐	☐	☐	☐	☐	☑	☐	☑
WiFi	☑	☑	☑	☑	☑	☑	☑	☑	☑	☑	☑	☑
Bluetooth	☑	☑	☑	☑	☐	☑	☑	☑	☑	☑	☑	☑
Speaker	☑	☑	☑	☑	☐	☑	☑	☑	☑	☑	☑	☑
Audio In	☑	☑	☑	☑	☐	☑	☑	☑	☑	☑	☑	☑
Accelerometer	☑	☑	☑	☑	☑	☑	☑	☑	☑	☑	☑	☑
Magnetometer	☐	☐	☑	☑	☐	☐	☐	☐	☑	☑	☑	☑
Gyroscope	☐	☐	☐	☑	☐	☐	☐	☑	☐	☐	☑	☑
GPS	☐	☑	☑	☑	☐	☐	☐	☐	☐	☑	☐	☑
Proximity Sensor	☑	☑	☑	☑	☐	☐	☐	☐	☐	☐	☐	☐
Camera	☑	☑	☑	☑	☐	☐	☐	☑	☐	☐	☑	☑
Video	☐	☐	☑	☑	☐	☐	☐	☑	☐	☐	☑	☑
Vibration	☑	☑	☑	☑	☐	☐	☐	☐	☐	☐	☐	☐

Most of the examples in this book will be built as iPhone however depending on the availability of hardware the examples will run equally well on the iPod touch and iPad; the underlying code is equally applicable as we're dealing for the most part directly with that hardware.

Differences Between iPhone and iPad

The most striking, and obvious, difference between the iPhone and the iPad is screen size. The original iPhone screen has 480×320 pixel resolution at 163 pixels per inch. The iPhone 4 and 4th generation iPod touch Retina Displays have a resolution of 960×640 pixel at 326 pixels per inch. Meanwhile both generations of the iPad screen have 1024×768 pixel resolution at 132 pixels per inch. This difference will be the single most fundamental thing to affect the way you design your user interface on the two platforms. Attempting to treat the iPad as simply a rather oversized iPod touch or iPhone will lead to badly designed applications. The metaphors you use on the two different platforms

The increased screen size of the device means that you can develop desktop-sized applications, not just phone-sized applications, for the iPad platform. Although in doing so, a rethink of the user interface to adapt to multi-touch is needed. What works for the iPhone or the desktop, won't automatically work on an iPad. For example, Apple totally redesigned the user interface of the iWork suite when they moved it to the iPad. If you're intending to port a Mac OS X desktop application to the iPad you should do something similar.

 Interestingly there is now an option for iOS developers to port their iPhone and iPad projects directly to Mac OS X. The Chameleon Project *http://chameleonproject.org* is a drop in replacement for UIKit that runs on Mac OS X, allowing iOS applications to be run on the desktop with little modification, in some cases none.

Due to its size and function the iPad is immediately associated in our minds with other more familiar objects like a legal pad or a book. Holding the device triggers powerful associations with these items, and we're mentally willing to accept the iPad has a successor to these objects. This is simply not true for the iPhone; the device is physically too small.

The Human Interface Guidelines

Apple has become almost infamous for strict adherence to its Human Interface Guidelines. Designed to present users with "a consistent visual and behavioral experience across applications and the operating system" the interface guidelines mean that (most) applications running on the Mac OS X desktop have a consistent look and feel. With the arrival of the iPhone and the iPad, Apple had to draw up new sets of guidelines for their mobile platforms, radically different from the traditional desktop.

Even for developers who are skeptical about whether they really needed to strictly adhere to the guidelines (especially when Apple periodically steps outside them) the Human Interface Guidelines have remained a benchmark against which the user experience can be measured.

Copies of the Human Interface Guidelines for both the iPhone and the iPad are available for download from the App Store Resource Center (*http://developer.apple.com/iphone/appstore*).

I would recommend that you read the mobile Human Interface Guidelines carefully, if only because violating them could lead to your application being rejected by the review team during the App Store approval process.

However this book is not about how to design your user interface or manage your user experience. For the most part the examples I present in this book are simple view-based applications that could be equally written for the iPhone and iPod touch or the iPad. The user interface is only there to illustrate how to use the underlying hardware. This book is about how to use the collection of sensors in these mobile devices.

Device Orientation and the iPad

The slider button on the side of the iPad can, optionally, be used to lock the device's orientation. This means that if you want the screen to stay in portrait mode, it won't move when you turn it sideways if locked. However despite the presence of the rotation lock (and unlike the iPhone where many applications only supported Portrait mode) an iPad application is expected to support all orientations equally.

 Apple has this to say about iPad applications: "An application's interface should support all landscape and portrait orientations. This behavior differs slightly from the iPhone, where running in both portrait and landscape modes is not required."

To implement basic support for all interface orientations, you should implement the shouldAutorotateToInterfaceOrientation: method in all of your application's view controllers, returning YES for all orientations. Additionally, you should configure the auto-resizing mark property of your views inside Interface Builder so that they correctly respond to layout changes (i.e. rotation of the device).

Going beyond basic support

If you want to go beyond basic support for alternative orientations there is more work involved. Firstly for custom views, where the placement of subviews is critical to the UI and need to be precisely located, you should override the layoutSubviews method to add your custom layout code. However, you should override this method only if the autoresizing behaviors of the subviews are not what you desire.

When an orientation event occurs, the UIWindow class will work with the front-most UIViewController to adjust the current view. Therefore if you need to perform tasks before, during, or after completing device rotation you should use the relevant rotation UIViewController notification methods. Specifically the view controller's willRotate ToInterfaceOrientation:duration:, willAnimateRotationToInterfaceOrienta tion:duration:, and didRotateFromInterfaceOrientation: methods are called at relevant points during rotation allowing you to perform tasks relevant to the orientation change in progress. For instance you might make use of these callbacks to allow you to add or remove specific views and reload your data in those views.

Detecting Hardware Differences

Because your application will likely support multiple devices, you'll need to write code to check which features are supported and adjust your application's behavior as appropriate.

Camera Availability

We cover the camera in detail in Chapter 2, however it is simple matter to determine whether a camera is present in the device:

```
BOOL available = [UIImagePickerController
    isSourceTypeAvailable:UIImagePickerControllerSourceTypeCamera];
```

Once you have determined that a camera is present you can enquire whether it supports video by making a call to determine the available media types the camera supports:

```
NSArray *media = [UIImagePickerController availableMediaTypesForSourceType:
                    UIImagePickerControllerSourceTypeCamera];
```

If the kUTTypeMovie media type is returned as part of the array, then the camera will support video recording:

```
if ( [media containsObject:(NSString *)kUTTypeMovie ] ){
    NSLog(@"Camera supports movie capture.");
}
```

Audio Input Availability

An initial poll of whether audio input is available can be done using the AVAudioSession class by checking the inputIsAvailable class property:

```
AVAudioSession *audioSession = [AVAudioSession sharedInstance];
BOOL audioAvailable = audioSession.inputIsAvailable;
```

 You will need to add the *AVFoundation.Framework* (right-click/Control-click on the Frameworks folder in Xcode, then choose Add→Existing Frameworks). You'll also need to import the header (put this in your declaration if you plan to implement the AVAudioSessionDelegate protocol discussed later):

```
#import <AVFoundation/AVFoundation.h>
```

You can also be notified of any changes in the availability of audio input, e.g., if a second generation iPod touch user has plugged in headphones with microphone capabilities. First, nominate your class as a delegate:

```
audioSession.delegate = self;
```

And then declare it as implementing the AVAudioSessionDelegate protocol in the declaration:

```
@interface YourAppDelegate : NSObject <UIApplicationDelegate,
    AVAudioSessionDelegate >
```

Then implement the inputIsAvailableChanged: in the implementation:

```
- (void)inputIsAvailableChanged:(BOOL)audioAvailable {
    NSLog(@"Audio availability has changed");
}
```

GPS Availability

The short answer to a commonly asked question is that the Core Location framework does not provide any way to get direct information about the availability of specific hardware such as the GPS at application run time, although you can check whether location services are enabled:

```
BOOL locationAvailable = [CLLocationManager locationServicesEnabled];
```

However, you can require the presence of GPS hardware for your application to load (see "Setting Required Hardware Capabilities").

Magnetometer Availability

Fortunately Core Location does allow you to check for the presence of the magneto-meter (digital compass) fairly simply:

```
BOOL magnetometerAvailable = [[CLLocationManager headingAvailable];
```

Setting Required Hardware Capabilities

If your application requires specific hardware features in order to run you can add a list of required capabilities to your application's *Info.plist* file. Your application will not start unless those capabilities are present on the device.

To do this, open the project and click on the application's *Info.plist* file to open it in the Xcode editor. Click on the bottommost entry in the list. A plus button will appear to the right-hand side of the key-value pair table.

Click on this button to add a new row to the table, and scroll down the list of possible options and select "Required device capabilities" (the UIRequiredDeviceCapabilities key). This will add an (empty) array to the *plist* file.

The allowed values for the keys are:

* telephony
* wifi
* sms
* still-camera
* auto-focus-camera
* front-facing-camera
* camera-flash

- video-camera
- accelerometer
- gyroscope
- location-services
- gps
- magnetometer
- gamekit
- microphone
- opengles-1
- opengles-2
- armv6
- armv7
- peer-peer

A full description of the possible keys is given in the Device Support section of the iPhone Application Programming Guide available from the iPhone Development Center.

Persistent WiFi

If your application requires a persistent WiFi connection you can set the Boolean UIRequiresPersistentWiFi key in the Application's *Info.plist* file to ensure that WiFi is available. If set to YES the operating system will open a WiFi connection when your application is launched and keep it open while the application is running. If this key is not present, or is set to NO, the Operating System will close the active WiFi connection after 30 minutes.

Background Modes

Setting the UIBackgroundModes key in the Application's *Info.plist* file notifies the operating systems that the application should continue to run in the background, after the user closes it, since it provides specific background services.

 Apple has this to say about background modes, "These keys should be used sparingly and only by applications providing the indicated services. Where alternatives for running in the background exist, those alternatives should be used instead. For example, applications can use the significant location change interface to receive location events instead of registering as a background location application."

There are three possible key values: `audio`, `location`, and `voip`. The `audio` key indicates that after closing the application will continue to play audible content. The `location` key indicates that the application provides location-based information for the user using the standard Core Location services, rather than the newer significant location change service. Finally, the `voip` key indicates that the application provides Voice-over-IP services. Applications marked with this key are automatically launched after system boot so that the application can attempt to re-establish VoIP services.

CHAPTER 2

Using the Camera

Phones with cameras only started appearing on the market in late 2001; now they're everywhere. By the end of 2003 more camera phones were sold worldwide than stand-alone digital cameras, and by 2006 half of the world's mobile phones had a built-in camera.

The social impact of this phenomenon should not be underestimated; the ubiquity of these devices has had a profound affect on society and on the way that news and information propagate. Mobile phones are constantly carried, which means their camera is always available. This constant availability has led to some innovative third party applications, especially with the new generation of smart phones. The iPhone has been designed with always-on connectivity in mind.

The Hardware

Until recently, only the iPhone has featured a camera in all of the available models. However the latest generation of both the iPod touch and iPad now also have cameras.

The original iPhone and iPhone 3G feature a fixed-focus 2.0-megapixel camera, while the iPhone 3GS features a 3.2-megapixel camera with auto-focus, auto-white balance and auto-macro focus (up to 10cm). The iPhone 3GS camera is also able of capturing 640×480 pixel video at 30 frames per second. Although the earlier models are physically capable of capturing video, they are limited in software and this feature is not available at the user level. The latest iPhone 4 features a 5-megapixel camera with better low-light sensitivity and backside illuminated sensor. The camera has an LED flash and is capable of capturing 720p HD video at 30 frames per second. The iPhone 4 also has a lower-resolution front-facing camera, which is capable of capturing 360p HD video at 30 frames per second.

The iPhone 3GS and iPhone 4 cameras are known to suffer from *rolling shutter effect* when used to take video. This effect is a form of aliasing that may result in distortion of fast moving objects, or image effects due to lighting levels that change as a frame is captured. At the time of writing it's not clear whether the 4th generation iPod touch and iPad 2 cameras suffer the same problem.

The latest generation of iPod touch and iPad also have both rear- and front-facing cameras, both of which are far lower resolution than the camera fitted to the iPhone 4, see Table 2-1 for details. You'll notice the difference in sizes between still and video images on the iPod touch and the iPad 2. It's unclear whether Apple is using a 1280×720 sensor and cropping off the left and right sides of the video image for still images, or whether it is using a 960×720 sensor and up-scaling it on the sides for video. The later would be an unusual approach for Apple, but is not inconceivable.

Table 2-1. Camera hardware support in various iPhone models

Model	Focus	Flash	Megapixels	Size	Video
Original iPhone	Fixed	No	2.0	1600×1200	No
iPhone 3G	Fixed	No	2.0	1600×1200	No
iPhone 3GS	Autofocus	No	3.2	2048×1536	VGA at 30fps
iPhone 4	Autofocus	LED flash	5.0 for still	2592×1944	720p at 30fps
			1.4 for video	1280×1024	
	Fixed	No	1.4	1280×1024	360p at 30fps
iPod touch (4th Gen)	Fixed	No	0.69 for still	960×720	720p at 30fps
			0.92 for video	1280×720	
	Fixed	No	1.4	1280×1024	VGA at 30fps
iPad 2	Fixed	No	0.69 for still	960×720	720p at 30fps
			0.92 for video	1280×720	
	Fixed	No	1.4	1280×1024	VGA at 30fps

All models produce geocoded images by default.

Capturing Stills and Video

The `UIImagePickerViewController` is an Apple-supplied interface for choosing images and movies, and taking new images or movies (on supported devices). This class handles all of the required interaction with the user and is very simple to use. All you need to do is tell it to start, then dismiss it after the user selects an image or movie.

Let's go ahead and build a simple application to illustrate how to use the image picker controller. Open Xcode and start a new project. Select a View-based Application for the iPhone, and name it *Media* when requested.

The first thing to do is set up the main view. This is going to consist of a single button that is pressed to bring up the Image Picker controller. An UIImageView will display the image, or thumbnail of the video, that is captured.

Select the *MediaViewController.h* interface file to open it in the editor and add a UIBut ton and an associated method to the interface file. Flag these as an IBOutlet and IBAction respectively. You also need to add a UIImageView to display that image returned by the image picker, which also needs to be flagged as an IBOutlet. Finally, add a UIImage PickerController, and flag the view controller as both UIImagePickerControllerDele gate and UINavigationControllerDelegate. The code to add to the default template is shown in bold:

```
#import <UIKit/UIKit.h>

@interface MediaViewController : UIViewController
    <UIImagePickerControllerDelegate, UINavigationControllerDelegate> {❶

    IBOutlet UIButton *pickButton;
    IBOutlet UIImageView *imageView;
    UIImagePickerController *pickerController;
}

-(IBAction) pickImage:(id) sender;

@end
```

❶ Both UIImagePickerControllerDelegate and UINavigationControllerDelegate dec-larations are necessary for the class to interact with the UIImagePickerController.

Next, open the *MediaViewController.m* implementation file and add a stub for the pickImage: method. As always, remember to release the pickButton, imageView and the pickerController in the dealloc method:

```
-(IBAction) pickImage:(id) sender {
    // Code goes here later
}

- (void)dealloc {
    [pickButton release];
    [imageView release];
    [pickerController release];
    [super dealloc];
}
```

After saving your changes (⌘-S) single click on the *MediaViewController.xib* NIB file to open it in Interface Builder. Drag and drop a UIButton and a UIImageView into the main View window. Go ahead and change the button text to something appropriate, and in the *Attributes Inspector* of the Utilities panel set the UIImageView's view mode to be `Aspect Fit`. Use the Size inspector resize the `UIImageView` to a 4:3 ratio. I used 280×210 points which fits nicely in a Portrait-mode iPhone screen.

Next click on "File's Owner" in the main panel. In the Connections inspector of the Utilities panel, connect both the `pickButton` outlet and the `pickImage:` received action to the button you just dropped into the View choosing Touch Up Inside as the action, see Figure 2-1.

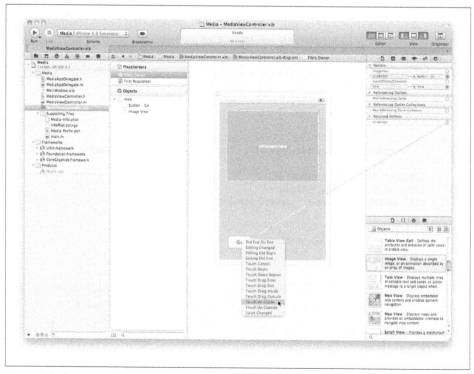

Figure 2-1. Connecting the UIButton to File's Owner

Then connect the `imageView` outlet to the `UIImageView` in our user interface.

Click on the *MediaViewController.m* implementation file and uncomment the `viewDid Load:` method. You're going to use this to initialize the `UIImagePickerController`. Make the changes shown in bold:

```
- (void)viewDidLoad {
    [super viewDidLoad];
    pickerController = [[UIImagePickerController alloc] init];❶
    pickerController.allowsEditing = NO;❷
    pickerController.delegate = self;❸
}
```

❶ This allocates and initializes the `UIImagePickerController`; don't forget to release it inside the `dealloc` method.

❷ This line prevents the picker controller from displaying the crop and resize tools. If enabled, the "crop and resize" stage is shown after capturing a still. For video, the trimming interface is presented.

❸ This line sets the delegate class to be the current class, the `MediaViewController`.

The `UIImagePickerController` can be directed to select an image (or video) from three image sources: `UIImagePickerControllerSourceTypeCamera`, `UIImagePickerController SourceTypePhotoLibrary` and `UIImagePickerControllerSourceTypeSavedPhotosAlbum`. Each presents a different view to the user allowing her to take an image (or a video) with the camera, from the image library, or from the saved photo album.

Now write the `pickImage:` method that will present the image picker controller to the user. There are a few good ways to do that, depending on the interface you want to present. The first method, makes use of a `UIActionSheet` to choose the source type, presenting the user with a list to decide whether they will take a still image or a video:

```
-(void)pickImage: (id)sender {
    UIActionSheet *popupQuery = [[UIActionSheet alloc]
        initWithTitle:nil
        delegate:self
        cancelButtonTitle:@"Cancel"
        destructiveButtonTitle:nil
        otherButtonTitles:@"Photo",@"Video",nil];

    popupQuery.actionSheetStyle = UIActionSheetStyleBlackOpaque;
    [popupQuery showInView:self.view];
    [popupQuery release];
}
```

If we're going to use this method we must specify that the view controller supports the `UIActionSheetDelegate` protocol in the interface file:

```
@interface MediaViewController : UIViewController
    <UIImagePickerControllerDelegate, UINavigationControllerDelegate,
    UIActionSheetDelegate> {
```

In the implementation file, provide an `actionSheet:clickedButtonAtIndex:` delegate method to handle presenting the image picker interface. If there is no camera present the source will be set to the saved photos album:

```
- (void)actionSheet:(UIActionSheet *)actionSheet
    clickedButtonAtIndex:(NSInteger)buttonIndex {

    if ([UIImagePickerController
        isSourceTypeAvailable:UIImagePickerControllerSourceTypeCamera]) {❶

        pickerController.sourceType = UIImagePickerControllerSourceTypeCamera;
    } else {❷
        pickerController.sourceType =
        UIImagePickerControllerSourceTypeSavedPhotosAlbum;
```

```
        }

        if (buttonIndex == 0) {
            pickerController.mediaTypes = [NSArray arrayWithObject: kUTTypeImage];
        } else if (buttonIndex == 1) {
            pickerController.mediaTypes = [NSArray arrayWithObject: kUTTypeMovie];
        }
        [self presentModalViewController:pickerController animated:YES];
    }
```

❶ Here we check whether the camera is available; if it is we set the sourceType to be the camera.

❷ If the camera is not available, we set the sourceType to be the Saved Photo Album.

Since we've made use of the kUTTypeImage and kUTTypeMovie type codes in this method we have to add the Mobile Core Services framework to our project.

 For those of you used to working in Xcode 3, the way you add frameworks to your project has changed. In the past you were able to right-click on the Framework's group and then select Add→Existing Frameworks. Unfortunately this is no longer possible and adding frameworks has become a more laborious process.

To add the framework, select the Media project file in the Project navigator window. You should see a panel as in see Figure 2-2. Select the Target and click on the Build Phases tab. Select the Link with Libraries drop down and use the + button to add the *MobileCoreServices.framework* from the list of available frameworks.

Add the following to the view controller interface file:

```
#import <MobileCoreServices/MobileCoreServices.h>
```

After saving the changes you can click on the Build and Run button. You should be presented with an interface much like Figure 2-3 (left). Clicking on the "Go" button you should be presented with the UIActionSheet that prompts the user to choose between still image and video capture.

 If you do go ahead and test the application in the iPhone Simulator you'll notice that there aren't any images in the Saved Photos folder, see Figure 2-3 (right). However there is a way around this problem. In the Simulator, tap on the Safari Icon and drag and drop a picture from your computer (you can drag it from the Finder or iPhoto) into the browser. From the browser you can save the image to the Saved Photos folder.

Instead of explicitly choosing an image or video via the action sheet, you could instead allow the user to pick the source. The following alternative code determines whether your device supports a camera and adds all of the available media types to an array. If there is no camera present the source will again be set to the saved photos album:

```
-(void)pickImage: (id)sender {
    if ([UIImagePickerController
            isSourceTypeAvailable:UIImagePickerControllerSourceTypeCamera]) {
        pickerController.sourceType = UIImagePickerControllerSourceTypeCamera;
        NSArray* mediaTypes =
            [UIImagePickerController availableMediaTypesForSourceType:
                               UIImagePickerControllerSourceTypeCamera];
        pickerController.mediaTypes = mediaTypes;
    } else {
        pickerController.sourceType =
            UIImagePickerControllerSourceTypeSavedPhotosAlbum;
    }
    [self presentModalViewController:pickerController animated:YES];
}
```

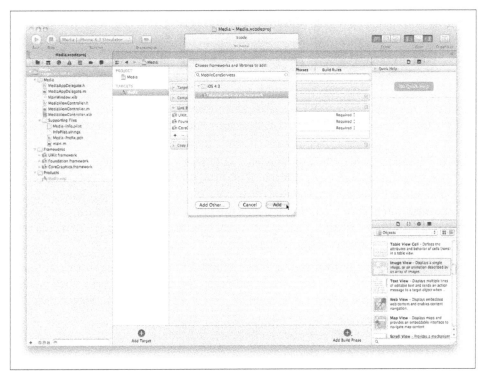

Figure 2-2. Adding the MobileCoreServices.framework to the project

Here instead of presenting an action sheet and allowing the user to choose which source type they wish to use we interrogate the hardware and decide which source types are available. We can see the different interfaces these two methods generate in Figure 2-4. The left interface is the still camera interface, the middle image is the video camera interface and the final (right-hand) image is the joint interface, which allows the user to either take still image or video.

Figure 2-3. The initial Media application interface (left), the UIActionSheet that pops up when the button is pressed (middle), and the Saved Photos folder that appears when the camera is unavailable (right)

The final interface, where the user may choose to return either a still image or a video, is the one presented by the second version of the pickImage: method. This code is also more flexible as it will run unmodified on any of the iPhone models that have a camera device. If your application requires either a still image or a video (and can not handle both) you should be careful to specify either kUTTypeImage or kUTTypeMovie media type as you did in the first version of the method.

You can choose either of the two different methods I've talked about above to present the image picker controller to the user. In either case when the user has finished picking an image (or video) the following delegate method will be called in the view controller:

```
-(void)imagePickerController:(UIImagePickerController *)picker
    didFinishPickingMediaWithInfo:(NSDictionary *)info {

  if( [info objectForKey:@"UIImagePickerControllerMediaType"] ==
    kUTTypeMovie ) {

    // add code here

  } else {
    imageView.image =
      [info objectForKey:@"UIImagePickerControllerOriginalImage"];

  }
  [self dismissModalViewControllerAnimated:YES];❶
}
```

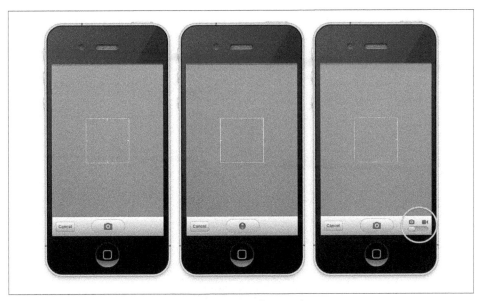

Figure 2-4. The three different UIImagePickerController interfaces

❶ We must dismiss the image picker interface in all cases.

When the `UIImagePickerController` returns it passes an `NSDictionary` containing a number of keys, listed in Table 2-2. Use the `UIImagePickerControllerMediaType` key to decide whether the image picker is returning a still image or a movie to its delegate method.

Table 2-2. Keys from the NSDictionary returned by the image picker controller

Key	Object type
UIImagePickerControllerMediaType	kUTTypeImage or kUTTypeMovie
UIImagePickerControllerOriginalImage	UIImage
UIImagePickerControllerEditedImage	UIImage
UIImagePickerControllerCropRect	CGRect
UIImagePickerControllerMediaURL	NSURL

We can retrieve the original image (or cropped version if editing is enabled) directly from the `NSDictionary` that was passed into the delegate method. This image reference can be passed directly to the `UIImageView` and displayed, as shown in the code in the next section and Figure 2-5.

Figure 2-5. A thumbnail of a single frame of video displayed in a UIImageView

Video Thumbnails

There is no easy way to retrieve a thumbnail of a video, unlike still photos. This section illustrates two methods of grabbing raw image data from an image picker.

Video Thumbnails Using the UIImagePicker

One way to grab a video frame for creating a thumbnail is to drop down to the underlying Quartz framework to capture an image of the picker itself. To do so, add the following highlighted code to the image picker delegate described previously in this chapter:

```
-(void)imagePickerController:(UIImagePickerController *)picker
    didFinishPickingMediaWithInfo:(NSDictionary *)info {

    if( [info objectForKey:@"UIImagePickerControllerMediaType"] ==
        kUTTypeMovie ) {
        CGSize pickerSize = CGSizeMake(picker.view.bounds.size.width,
                                        picker.view.bounds.size.height-100);
        UIGraphicsBeginImageContext(pickerSize);
        [picker.view.layer renderInContext:UIGraphicsGetCurrentContext()];
        UIImage *thumbnail = UIGraphicsGetImageFromCurrentImageContext();
        UIGraphicsEndImageContext();
        imageView.image = thumbnail;
```

```
    } else {
        imageView.image = image;
    }

    }
    [self dismissModalViewControllerAnimated:YES];
}
```

Since `picker.view.layer` is part of the `UIView` parent class and is of type `CALayer`, the compiler doesn't know about `renderInContext:` method unless you import the Quartz-Core header file. Add the following to the implementation file:

```
#import <QuartzCore/QuartzCore.h>
```

Video Thumbnails Using AVFoundation

Another method to obtain a thumbnail that will result in a better image is to use the `AVFoundation` framework. First replace the code you added in the previous section with the highlighted code below:

```
-(void)imagePickerController:(UIImagePickerController *)picker
        didFinishPickingMediaWithInfo:(NSDictionary *)info {

    if( [info objectForKey:@"UIImagePickerControllerMediaType"] ==
    kUTTypeMovie ) {

        AVURLAsset *asset=[[AVURLAsset alloc]
            initWithURL:[info objectForKey:UIImagePickerControllerMediaURL]
                               options:nil];
        AVAssetImageGenerator *generator =
            [[AVAssetImageGenerator alloc] initWithAsset:asset];
        generator.appliesPreferredTrackTransform=TRUE;
        [asset release];
        CMTime thumbTime = CMTimeMakeWithSeconds(0,30);

        AVAssetImageGeneratorCompletionHandler handler =
             ^(CMTime requestedTime, CGImageRef im, CMTime actualTime,
               AVAssetImageGeneratorResult result, NSError *error) {
            if (result != AVAssetImageGeneratorSucceeded) {
                NSLog(@"Error:%@", error);
            }
            imageView.image = [[UIImage imageWithCGImage:im] retain];
            [generator release];
        };

        CGSize maxSize = CGSizeMake(320, 180);
        generator.maximumSize = maxSize;
        [generator generateCGImagesAsynchronouslyForTimes:
            [NSArray arrayWithObject:[NSValue valueWithCMTime:thumbTime]]
                    completionHandler:handler];

    } else {
        imageView.image = image;
```

```
        }
        [self dismissModalViewControllerAnimated:YES];
    }
```

Then make sure to add the AVFoundation and CoreMedia frameworks to the project by importing the header files at the top of the implementation:

```
#import <AVFoundation/AVFoundation.h>
#import <CoreMedia/CoreMedia.h>
```

The only real downside of this method is that AVAssetImageGenerator makes use of key frames, which are typically spaced at one second intervals. Hopefully the key frame will make a good thumbnail image.

Saving Media to the Photo Album

You can save both images and videos to the Photo Album using the UIImageWriteToSa vedPhotosAlbum and UISaveVideoAtPathToSavedPhotosAlbum methods. The method will also obtain a thumbnail image for the video if desired.

The saving functions in this example are asynchronous; if the application is interrupted (e.g., takes a phone call) or terminated, the image or video will be lost. You need to ensure that your user is aware that processing is happening in the background as part of your application interface.

The following example save the image to the Photo Album by adding a call to UIImage WriteToSavedPhotosAlbum to the image picker delegate. The example will then provide feedback when the image has been successfully saved or an error occurs. Add the following highlighted lines to the image picker controller presented earlier in the chapter:

```
-(void)imagePickerController:(UIImagePickerController *)picker
        didFinishPickingMediaWithInfo:(NSDictionary *)info {

    if( [info objectForKey:@"UIImagePickerControllerMediaType"] ==
        kUTTypeMovie ) {

        CGSize pickerSize = CGSizeMake(picker.view.bounds.size.width,
                                       picker.view.bounds.size.height-100);
        UIGraphicsBeginImageContext(pickerSize);
        [picker.view.layer renderInContext:UIGraphicsGetCurrentContext()];
        UIImage *thumbnail = UIGraphicsGetImageFromCurrentImageContext();
        UIGraphicsEndImageContext();
        imageView.image = thumbnail;

    } else {
        UIImage *image =
            [info objectForKey:@"UIImagePickerControllerOriginalImage"];
        UIImageWriteToSavedPhotosAlbum(
            image,
            self,
            @selector(
              imageSavedToPhotosAlbum:didFinishSavingWithError:contextInfo:),
```

```
            nil);

        imageView.image = image;
    }
    [self dismissModalViewControllerAnimated:YES];
}
```

Then add the following method, which presents a UIAlertView notifying the user that
the save has occurred:

```
- (void)imageSavedToPhotosAlbum:(UIImage *)image
      didFinishSavingWithError:(NSError *)error
      contextInfo:(void *)contextInfo {

   NSString *title;
   NSString *message;
   if (!error) {
       title = @"Photo Saved";
       message = @"The photo has been saved to your Photo Album";
   } else {
       title = NSLocalizedString(@"Error Saving Photo", @"");
       message = [error description];
   }
   UIAlertView *alert = [[UIAlertView alloc] initWithTitle:title
                                            message:message
                                            delegate:nil
                                    cancelButtonTitle:@"OK"
                                    otherButtonTitles:nil];

   [alert show];
   [alert release];
}
```

 The call to UIImageWriteToSavedPhotosAlbum can typically take up to 4
seconds to complete in the background. If the application is interrupted
or terminated during this time then the image may not have been saved.

You can similarly add the following highlighted lines to the delegate method to save
captured video:

```
-(void)imagePickerController:(UIImagePickerController *)picker
      didFinishPickingMediaWithInfo:(NSDictionary *)info {

  if( [info objectForKey:@"UIImagePickerControllerMediaType"] == kUTTypeMovie ) {
     NSString *tempFilePath =
        [[info objectForKey:UIImagePickerControllerMediaURL] path];
     if ( UIVideoAtPathIsCompatibleWithSavedPhotosAlbum(tempFilePath) ) {
       UISaveVideoAtPathToSavedPhotosAlbum( tempFilePath, self,
          @selector(video:didFinishSavingWithError:contextInfo:),
          tempFilePath);
     }

     CGSize pickerSize = CGSizeMake(picker.view.bounds.size.width,
                                    picker.view.bounds.size.height-100);
     UIGraphicsBeginImageContext(pickerSize);
```

```
        [picker.view.layer renderInContext:UIGraphicsGetCurrentContext()];
        UIImage *thumbnail = UIGraphicsGetImageFromCurrentImageContext();
        UIGraphicsEndImageContext();
        imageView.image = thumbnail;

    } else {
        UIImage *image =
          [info objectForKey:@"UIImagePickerControllerOriginalImage"];
        UIImageWriteToSavedPhotosAlbum(image, self,
            @selector(
              imageSavedToPhotosAlbum:didFinishSavingWithError:contextInfo:), nil);

        imageView.image = image;
    }
    [self dismissModalViewControllerAnimated:YES];
}
```

Next add the following method to report whether the video has been successfully saved to the device's Photo Album, or an error occurred:

```
- (void)video:(NSString *)videoPath
      didFinishSavingWithError:(NSError *)error
      contextInfo:(NSString *)contextInfo {

    NSString *title;
    NSString *message;
    if (!error) {
        title = @"Video Saved";
        message = @"The video has been saved to your Photo Album";
    } else {
        title = NSLocalizedString(@"Error Saving Video", @"");
        message = [error description];
    }
    UIAlertView *alert = [[UIAlertView alloc] initWithTitle:title
                                                    message:message
                                                   delegate:nil
                                          cancelButtonTitle:@"OK"
                                          otherButtonTitles:nil];
    [alert show];
    [alert release];
}
```

Make sure you've saved your changes, and click on the Run button in the Xcode toolbar to compile and deploy the application to your device. If everything is working, you will see a thumbnail after you take a photo or video. After a few seconds a confirmation dialog will appear reporting success or an error. See Figure 2-6.

Figure 2-6. Saving images (or video) to the Photo Album

Video Customization

If you are capturing video you can make some video-specific customizations using the `videoQuality` and `videoMaximumDuration` properties of the `UIImagePickerController` class:

```
pickerController.videoQuality = UIImagePickerControllerQualityTypeLow;
pickerController.videoMaximumDuration = 90; // Maximum 90 seconds duration
```

Table 2-3 illustrates the expected sizes of a typical 90 second movie file for the three possible image quality levels, which defaults to `UIImagePickerControllerQualityType Medium`.

Table 2-3. Size of 90 seconds duration video for different quality settings

Quality	Size
UIImagePickerControllerQualityTypeLow	1.8 MB
UIImagePickerControllerQualityTypeMedium	8.4 MB
UIImagePickerControllerQualityTypeHigh	32 MB

The maximum, and default, value for the `videoMaximumDuration` property is 10 minutes. Users are forced to trim longer video to match the duration you request.

CHAPTER 3

Using Audio

The main classes for handling audio in the SDK are in the AVFoundation and Media Player frameworks. This chapter will provide a brief overview of how to play and record audio using these frameworks.

The Hardware

Whilst most phones have only one microphone, iPhone 4 has two. The main microphone is located normally on the bottom next to the dock connector, while the second microphone is built into the top near the headphone jack. This second microphone is intended for video-calling, but is also used in conjunction with the main microphone to suppress background noise.

In comparison the iPad 2 has a single microphone, but there is a difference between the two models which could lead to a difference in audio recording quality between the 3G and WiFi-only models. On the WiFi-only model, the microphone hole is built-into the back of the device, whereas on 3G models, it's built into the antenna casing. There are suggestions that this difference may lead to cleaner audio recordings with the WiFi model, with the 3G model sounding muffled and echo-prone by comparison.

Both the iPhone 4 and the iPad use an Apple branded Cirrus Logic 338S0589 for their audio DAC, with a frequency response of 20Hz to 20kHz, and audio sampling of 16-bit at 44.1kHz.

All of the current iPhone, iPad and iPod touch models use a 2.5mm 4-pole TRRS (tip, ring, ring, sleeve) connector which has a somewhat unorthodox mapping to the standard RCA connector as shown in Table 3-1.

Table 3-1. Mapping between the iPhone's 4-pole jack and the standard RCA connector colors

Apple	RCA
Tip	RCA White
1st Ring	RCA Yellow

Apple	RCA
2nd Ring	RCA Ground
Sleeve	RCA Red

Media Playback

Let's first look at playing back existing media stored in the iPod library. Apple has provided convenience classes that allow you to select and play back iPod media inside your own application as part of the Media Player framework.

 The following examples make use of the iPod library; this is not present in the iPhone Simulator and will only work correctly on the device itself.

The approach uses picker controllers and delegates as in the previous chapter. In this example I use an MPMediaPickerController that, via the MPMediaPickerControllerDele gate protocol, returns an MPMediaItemCollection object containing the media items the user has selected. The collection of items can be played using an MPMusicPlayerControl ler object.

Lets go ahead and build a simple media player application to illustrate how to use the media picker controller. Open Xcode and start a new View-based Application project, naming it "Audio" when requested. Click on the Audio project file in the Project navigator window, select the Target and click on the Build Phases tab. Click on the Link with Libraries drop down and click on the + button to add the MediaPlayer framework.

Edit the *AudioViewController.h* interface file to import the MediaPlayer framework and declare the class as an MPMediaPickerControllerDelegate. Then add the IBOutlet instance variables and IBAction methods for the buttons we will create in Interface Builder:

```
#import <UIKit/UIKit.h>
#import <MediaPlayer/MediaPlayer.h>

@interface AudioViewController : UIViewController
  <MPMediaPickerControllerDelegate> {

    IBOutlet UIButton *pickButton;
    IBOutlet UIButton *playButton;
    IBOutlet UIButton *pauseButton;
    IBOutlet UIButton *stopButton;

    MPMusicPlayerController *musicPlayer;

}

- (IBAction)pushedPick:(id)sender;
- (IBAction)pushedPlay:(id)sender;
```

```
- (IBAction)pushedPause:(id)sender;
- (IBAction)pushedStop:(id)sender;
```

@end

Save your changes, and open the *AudioViewController.m* implementation file. In the pushedPick: method, instantiate an MPMediaPickerController object. The view will be modal, which means that the user must make a selection to leave picking mode. We'll link this method directly to a button in the user interface:

```
-(IBAction) pushedPick:(id)sender {
    MPMediaPickerController *mediaPicker =
      [[MPMediaPickerController alloc]
        initWithMediaTypes: MPMediaTypeAnyAudio];
    mediaPicker.delegate = self;
    mediaPicker.allowsPickingMultipleItems = YES;
    [self presentModalViewController:mediaPicker animated:YES];
    [mediaPicker release];
}
```

You must now implement the following two delegate methods, which are used to dismiss the view controller and handle the returned items:

```
- (void) mediaPicker:(MPMediaPickerController *)
       mediaPicker didPickMediaItems:(MPMediaItemCollection *)
       userMediaItemCollection {

    [self dismissModalViewControllerAnimated: YES];

    musicPlayer = [MPMusicPlayerController applicationMusicPlayer];❶
    [musicPlayer setQueueWithItemCollection: userMediaItemCollection];
}

- (void) mediaPickerDidCancel: (MPMediaPickerController *) mediaPicker {
    [self dismissModalViewControllerAnimated: YES];
}
```

❶ The MPMusicPlayerController responds to all the messages you might expect, e.g., play, pause, stop, volume, etc. These can be linked to buttons in the user interface if you want to give users direct control over these functions.

You'll link the remaining methods directly to buttons in the user interface:

```
-(IBAction) pushedPlay:(id)sender {
    [musicPlayer play];
}

-(IBAction) pushedPause:(id)sender {
    [musicPlayer pause];
}

-(IBAction) pushedStop:(id)sender {
    [musicPlayer stop];
}
```

Remember to release the instance objects in the **dealloc** method:

```
- (void)dealloc {
    [pickButton release];
    [playButton release];
    [pauseButton release];
    [stopButton release];
    [musicPlayer release];
    [super dealloc];
}
```

Save your changes, and click on the *AudioViewController.xib* NIB file to open it in Interface Builder. Drag four **UIButton** elements from the Library window into the View window. Double click on each of them and change the default text to be "Pick", "Play", "Pause" and "Stop". Then open the Assistant Editor (View→Assistant Editor→Show Assistant Editor) and Control-Click and drag to associate the buttons with their respective **IBOutlet** and **IBAction** outlets and actions in the *AudioViewController.h* interface file, see Figure 3-1.

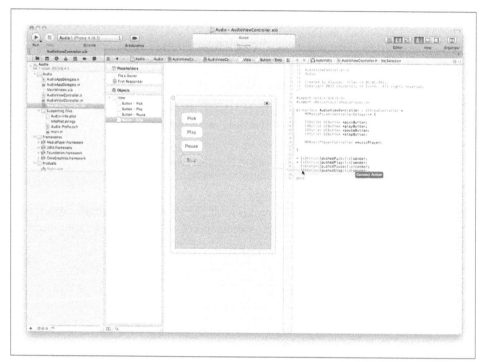

Figure 3-1. Connecting the application's user interface in Interface Builder

Save your changes and click on the Run button in the Xcode toolbar to build and deploy the code to your device.

Remember that you'll need to test the application on your device.

Once the application loads, tap on the "Pick" button to bring up the picker controller, select some songs, and tap the Done button.(see Figure 3-2). Press the "Play" button and the music you selected should start playing.

Figure 3-2. The initial main view (left), the MPMediaPickerController (middle), and the main interface while a song is being played (right)

Once playback has begun you need to keep track of the currently playing item and display that to the user. At the very least you must provide some way for the user to pause or stop playback, and perhaps to change their selection. The MPMusicPlayerCon troller class provides two methods: the beginGeneratingPlaybackNotifications: method, and a corresponding endGeneratingPlaybackNotifications:. Add the high-lighted line below to your mediaPicker:didPickMediaItems: delegate method:

```
- (void) mediaPicker:(MPMediaPickerController *) mediaPicker
      didPickMediaItems:(MPMediaItemCollection *) userMediaItemCollection {

    [self dismissModalViewControllerAnimated: YES];

    musicPlayer = [MPMusicPlayerController applicationMusicPlayer];
    [musicPlayer setQueueWithItemCollection: userMediaItemCollection];
```

```
        [musicPlayer beginGeneratingPlaybackNotifications];
    }
```

When the begin method is invoked the class will start to generate notifications when the player state changes and when the current playback item changes. Your application can access this information by adding itself as an observer using the NSNotification Center class:

```
    - (void) mediaPicker:(MPMediaPickerController *) mediaPicker
         didPickMediaItems:(MPMediaItemCollection *) userMediaItemCollection {

        [self dismissModalViewControllerAnimated: YES];

        musicPlayer = [MPMusicPlayerController applicationMusicPlayer];
        [musicPlayer setQueueWithItemCollection: userMediaItemCollection];
        [musicPlayer beginGeneratingPlaybackNotifications];

        NSNotificationCenter *notificationCenter =
          [NSNotificationCenter defaultCenter];
        [notificationCenter addObserver:self
        selector:@selector(handleNowPlayingItemChanged:)
            name:@"MPMusicPlayerControllerNowPlayingItemDidChangeNotification"
          object:musicPlayer];

        [notificationCenter addObserver:self
         selector:@selector(handlePlaybackStateChanged:)
             name:@"MPMusicPlayerControllerPlaybackStateDidChangeNotification"
           object:musicPlayer];
    }
```

This will invoke the selector methods in the class when the appropriate notification arrives. You could, for example, use the first to update a UILabel in the view telling the user the name of the currently playing song.

For now let's just go ahead and implement these methods to print messages to the console log. In the *AudioViewController.m* implementation file, add the method below. This will be called when the current item being played changes:

```
    - (void)handleNowPlayingItemChanged:(id)notification {
        MPMediaItem *currentItem = [musicPlayer nowPlayingItem];❶
        NSString *title = [currentItem valueForProperty:MPMediaItemPropertyTitle];
        NSLog(@"Song title = %@", title);
    }
```

❶ Unusually, the MPMediaItem class only has one instance method, the valueForProp erty: method. This is because the class can wrap a number of media types, and each type can have a fairly wide range of metadata associated with it. A full list of possible keys can be found in the MPMediaItem class reference, but keys include MPMediaI temPropertyTitle, MPMediaItemPropertyArtwork, etc.

You can use this to update the user interface, e.g., changing the state of the play and stop buttons when the music ends:

```
- (void)handlePlaybackStateChanged:(id)notification {
    MPMusicPlaybackState playbackState = [musicPlayer playbackState];
    if (playbackState == MPMusicPlaybackStatePaused) {
        NSLog(@"Paused");

    } else if (playbackState == MPMusicPlaybackStatePlaying) {
        NSLog(@"Playing");

    } else if (playbackState == MPMusicPlaybackStateStopped) {
        NSLog(@"Stopped");

    }
}
```

Save your changes, and click on the Run button in the Xcode toolbar to build and deploy your code onto your device. Once your application loads, press the "Pick" button to bring up the pick controller again, select some songs, and press the "Done" button. Press "Play" and the music should start playing. You should also see something similar to the log messages below in the Debugger Console:

```
2011-06-01 19:23:07.602 Audio[2844:707] Song title = Affirmation
2011-06-01 19:23:07.617 Audio[2844:707] Playing
```

You could go on to develop the application by displaying information about the currently playing and queued songs. Let's move on from playing existing media and look at how to play and record your own audio on the device.

Recording and Playing Audio

The AVAudioRecorder class is part of the AVFoundation framework and provides audio recording capabilities for an application. The framework allows you to:

- Record until the user stops the recording
- Record for a specified duration
- Pause and resume a recording

The corresponding AVAudioPlayer class (also part of the AVFoundation framework) provides some fairly sophisticated functionality allowing you to play sound in your application. It can:

- Play sounds of any duration
- Play sounds from files or memory buffers
- Loop sounds
- Play multiple sounds simultaneously (one sound per audio player) with precise synchronization
- Control relative playback level and stereo positioning for each sound you are playing

- Seek to a particular point in a sound file, which supports such application features as fast forward and rewind

Recording Audio

Lets build a simple application to record some audio to a file and play it back later. Open Xcode and start a new View-based Application, naming it *Recorder* when requested.

When the Xcode project opens, add *both* the AVFoundation and CoreAudio frameworks to the project in a similar manner as we added the MediaPlayer framework to the Audio application earlier in the chapter.

Click on the *RecorderViewController.h* interface file to open it in the Standard Editor and make the following changes to the template file generated for you by Xcode:

```
#import <UIKit/UIKit.h>
#import <AVFoundation/AVFoundation.h>

@interface RecorderViewController : UIViewController
    <AVAudioRecorderDelegate> {

    IBOutlet UIButton *startStopButton;
    NSURL *tmpFile;
    AVAudioRecorder *recorder;
    BOOL recording;
}

- (IBAction)startStopButtonPressed;

@end
```

Save your changes and open the corresponding `RecorderViewController.xib` file in Interface Builder. Drag and drop a `UIButton` from the Object Library in the Utilities pane into the View and change the title text to "Start Recording". Then connect it to the `IBOutlet` and `IBAction` in your interface file using the Assistant Editor, as in Figure 3-3.

Save your changes and open the *RecorderViewController.m* implementation file in the Standard Editor, making the following changes to the default template generated by Xcode:

```
#import "RecorderViewController.h"
#import <CoreAudio/CoreAudioTypes.h>

@implementation RecorderViewController

- (IBAction)startStopButtonPressed {

    AVAudioSession * audioSession = [AVAudioSession sharedInstance];

    if (!recording) {

        // Add code here...
```

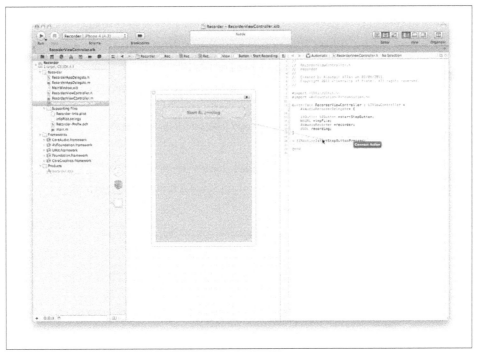

Figure 3-3. Connecting the user interface in Interface Builder

```
    } else {

            // Add code here...

    }

}

- (void)dealloc {
    [startStopButton release];
    [tmpFile release];
    [recorder release];
    [super dealloc];
}

- (void)didReceiveMemoryWarning {
    [super didReceiveMemoryWarning];

}

#pragma mark - View lifecycle

- (void)viewDidLoad {
    [super viewDidLoad];
    recording = NO;
```

```
}

- (void)viewDidUnload {
    [super viewDidUnload];
}

- (BOOL)shouldAutorotateToInterfaceOrientation:(UIInterfaceOrientation)
        interfaceOrientation {

    return (interfaceOrientation == UIInterfaceOrientationPortrait);
}

@end
```

Then make the following changes to the startStopButtonPressed method:

```
- (IBAction)startStopButtonPressed {

    AVAudioSession * audioSession = [AVAudioSession sharedInstance];

    if (!recording) {

        recording = YES;

        [audioSession setCategory:AVAudioSessionCategoryRecord error:nil];
        [audioSession setActive:YES error:nil];
        [startStopButton setTitle:@"Stop Recording"
                        forState:UIControlStateNormal];

        NSMutableDictionary* recordSetting =
          [[NSMutableDictionary alloc] init];
        [recordSetting setValue:
          [NSNumber numberWithInt:kAudioFormatAppleIMA4]
                        forKey:AVFormatIDKey];
        [recordSetting setValue:[NSNumber numberWithFloat:44100.0]
                        forKey:AVSampleRateKey];
        [recordSetting setValue:[NSNumber numberWithInt: 2]
                        forKey:AVNumberOfChannelsKey];

        tmpFile = [NSURL fileURLWithPath:
                        [NSTemporaryDirectory() stringByAppendingPathComponent:
                        [NSString stringWithFormat: @"%.0f.%@",
                          [NSDate timeIntervalSinceReferenceDate] * 1000.0,
                          @"caf"]]];

        recorder = [[AVAudioRecorder alloc] initWithURL:tmpFile
                                              settings:recordSetting
                                                 error:nil];
        [recorder setDelegate:self];
        [recorder prepareToRecord];
        [recorder record];

    } else {

        recording = NO;
        [audioSession setActive:NO error:nil];
```

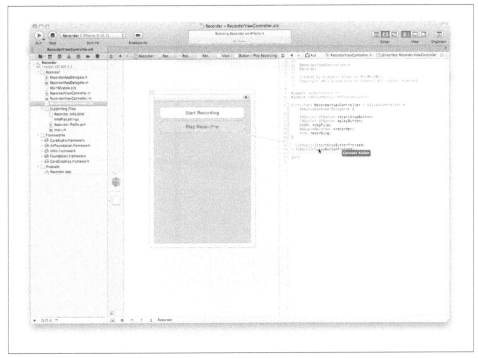

Figure 3-4. Adding the "Play Recording" button to the user interface

```
[startStopButton setTitle:@"Start Recording"
                  forState:UIControlStateNormal];
[recorder stop];

    }

}
```

If you save your changes and click on the Run button to build and deploy the application to your device you should see the "Start Recording" button changes to "Stop Recording" when pressed. Pressing the button again should change the text back to "Start Recording". In the next section I'll show you a way to check that the device is actually recording audio.

Playing Audio

Open up the *RecorderViewController.h* interface file and add the following IBOutlet instance variable:

```
IBOutlet UIButton *playButton;
```

along with the following IBAction method:

```
- (IBAction)playButtonPressed;
```

Figure 3-5. The finished Recorder application

Then single click on the *RecorderViewController.xib* file to open it in Interface Builder. Drag and drop and new UIButton into the view and change the title text to be "Play Recording". Use the Assistant Editor to connect the new button to the recently added IBOutlet and IBAction in the interface file, see Figure 3-4.

Save your changes, return to the *RecorderViewController.m* implementation file and add the following method implementation:

```
- (IBAction)playButtonPressed {
    AVAudioSession * audioSession = [AVAudioSession sharedInstance];
    [audioSession setCategory:AVAudioSessionCategoryPlayback error:nil];
    [audioSession setActive:YES error:nil];

    AVAudioPlayer * player =
      [[AVAudioPlayer alloc] initWithContentsOfURL:tmpFile error:nil];
    [player prepareToPlay];
    [player play];

}
```

Save your changes and click on the Run button in the Xcode toolbar to build and deploy the application to your device. You should see something like Figure 3-5.

If you now tap on the "Start Recording" button the title of the button should change to "Stop Recording"; speak into the iPhone's microphone for few seconds and tap the button again. Then tap on the "Play Recording" button and you should hear yourself speaking.

CHAPTER 4

Using the Accelerometer

An accelerometer measures the linear acceleration of the device. The original iPhone, and first generation iPod touch, use the LIS302DL 3-axis MEMS based accelerometer produced by STMicroelectronics. Later iPhone and iPod touch models use a similar LIS331DL chip, also manufactured by STMicroelectronics.

Both of these accelerometers can operate in two modes, allowing the chip to measure either ±2g and ±8g. In both modes the chip can sample at either 100 Mhz or 400 Mhz. Apple operates the accelerometer in the ±2g mode (presumably at 100 Mhz) with a nominal resolution of 0.018g. In the ±8g mode the resolution would be four times coarser, and the presumption must be that Apple decided better resolution would be more useful than a wider range. Under normal conditions the device will actually measure g-forces to approximately ±2.3g however measurements above a 2g are un-calibrated.

> While it should in theory be possible to change the operating mode of the accelerometer, there is currently no published API that allows you to do so within the SDK.

About the Accelerometer

The iPhone's accelerometer measures the linear acceleration of the device so it can report the device's roll and pitch, but not its yaw. If you are dealing with a device that has a digital compass you can combine the accelerometer and magnetometer readings to have roll, pitch, and yaw measurements (see Chapter 5 for details on how to access the magnetometer).

Yaw, *pitch*, and *roll* refer to the rotation of the device in three axes. If you think about an aircraft in the sky, pushing the nose down or pulling it up modifies the pitch angle of the aircraft. However, if you keep the nose straight ahead you can also modify the roll of the aircraft using the flaps; one wing will come up, the other will go down. By keeping the wings level you can use the tail flap to change the heading (or yaw) of the aircraft, rotating it in a 2D plane.

The accelerometer reports three figures: X, Y, and Z (see Figure 4-1). Acceleration values for each axis are reported directly by the hardware as G-force values. Therefore, a value of 1.0 represents a load of approximately 1-gravity (Earth's gravity). X corresponds to roll, Y to pitch, and Z to whether the device is front side up or front side down, with a value of 0.0 being reported when the iPhone is edge-on.

When dealing with acceleration measurements you must keep in mind that the accelerometer is measuring just that: the linear acceleration of the device. When at rest (in whatever orientation) the figures represent the force of gravity acting on the device, and correspond to the roll and pitch of the device (in the X and Y directions at least). But while in motion, the figures represent the acceleration due to gravity, plus the acceleration of the device itself relative to its rest frame.

Writing an Accelerometer Application

Let's go ahead and implement a simple application to illustrate how to approach the accelerometer. Open Xcode and start a new View-based application for the iPhone, and name the project "Accelerometer" when prompted for a filename.

 The raw accelerometer data can also be accessed using the Core Motion framework, which was new in iOS 4.0. I talk about how to do this in Chapter 6. It is therefore possible, even likely, that the UIAccelerome ter class discussed in this chapter my be deprecated in a future iOS release.

Click on the *AccelerometerViewController.xib* file to open it into Interface Builder. Since you want to both report the raw figures from the accelerometer and also display them using a progress bar, go ahead and drag and drop three UIProgressView controls from the Object Library into the View window. Then add two UILabel elements for each progress bar: one to hold the X, Y, or Z label and the other to hold the accelerometer measurements. After you do that, the view should look something a lot like Figure 4-2.

Go ahead and close the Utilities panel and click to open the Assistant Editor. Then Control-Click and drag from the three UIProgressView elements, and the three UILa bel elements to the *AcclerometerViewController.h* header file. The header file should

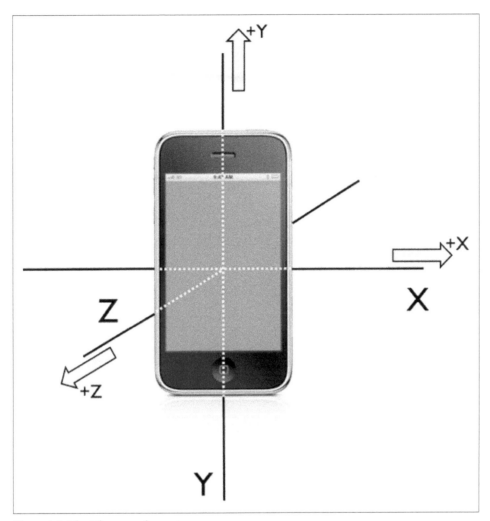

Figure 4-1. The iPhone accelerometer axes

be displayed in the Assistant Editor on the right-hand side of the Xcode 4 interface (see Figure 4-3).

This will automatically create and declare three `UILabel` and three `UIProgressView` variables as `IBOutlet` objects. Since they aren't going to be used outside the class, there isn't much point in declaring them as class properties, which you'd do by Control-click and drag from the element to outside the curly brace. After doing this the code should look like this:

```
#import <UIKit/UIKit.h>

@interface AccelerometerViewController : UIViewController {
```

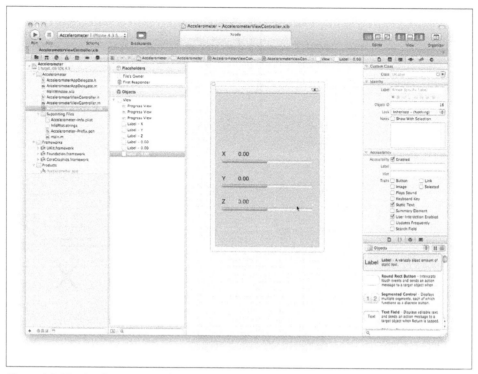

Figure 4-2. *The Accelerometer application UI*

```
    IBOutlet UIProgressView *xBar;
    IBOutlet UIProgressView *yBar;
    IBOutlet UIProgressView *zBar;

    IBOutlet UILabel *xLabel;
    IBOutlet UILabel *yLabel;
    IBOutlet UILabel *zLabel;
}

@end
```

Close the Assistant Editor, return to the Standard Editor and click on the *AccelerometerViewController.h* interface file. Now go ahead and set up a `UIAccelerometer` instance. Also declare the class as a `UIAccelerometerDelegate`. Here's how the should look when you are done:

```
#import <UIKit/UIKit.h>

@interface AccelerometerViewController :
   UIViewController <UIAccelerometerDelegate> { ❶

    IBOutlet UILabel *xLabel;
    IBOutlet UILabel *yLabel;
    IBOutlet UILabel *zLabel;
```

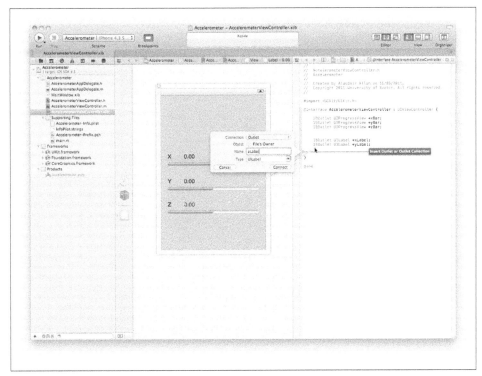

Figure 4-3. Connecting the UI elements to your code in Interface Builder

```
    IBOutlet UIProgressView *xBar;
    IBOutlet UIProgressView *yBar;
    IBOutlet UIProgressView *zBar;

    UIAccelerometer *accelerometer;
}

@end
```

❶ Here we declare that the class implements the UIAccelerometer delegate protocol.

Make sure you've saved your changes and click on the corresponding *Accelerometer-ViewController.m* implementation file to open it in the Xcode editor. You don't actually have to do very much here, as Interface Builder handled most of the heavy lifting by adding code to properly handle the user interface elements. Here's what the file should look like when you are done:

```
#import "AccelerometerViewController.h"

@implementation AccelerometerViewController

- (void)viewDidLoad {
```

```objc
    accelerometer = [UIAccelerometer sharedAccelerometer];❶
    accelerometer.updateInterval = 0.1;❷
    accelerometer.delegate = self;❸
    [super viewDidLoad];
}

- (void)didReceiveMemoryWarning {
    [super didReceiveMemoryWarning];
}

- (void)viewDidUnload
{
    [xBar release];
    xBar = nil;
    [yBar release];
    yBar = nil;
    [zBar release];
    zBar = nil;
    [xLabel release];
    xLabel = nil;
    [yLabel release];
    yLabel = nil;
    [zLabel release];
    zLabel = nil;
    [super viewDidUnload];
}

- (void)dealloc {
    [xLabel release];
    [yLabel release];
    [zLabel release];
    [xBar release];
    [yBar release];
    [zBar release];

    accelerometer.delegate = nil;
    [accelerometer release];

    [super dealloc];
}

- (BOOL)shouldAutorotateToInterfaceOrientation:
        (UIInterfaceOrientation)interfaceOrientation {

    return (interfaceOrientation == UIInterfaceOrientationPortrait);
}

#pragma mark UIAccelerometerDelegate Methods

- (void)accelerometer:(UIAccelerometer *)meter
  didAccelerate:(UIAcceleration *)acceleration ❹
{
    xLabel.text = [NSString stringWithFormat:@"%f", acceleration.x];
    xBar.progress = ABS(acceleration.x);
```

```
    yLabel.text = [NSString stringWithFormat:@"%f", acceleration.y];
    yBar.progress = ABS(acceleration.y);

    zLabel.text = [NSString stringWithFormat:@"%f", acceleration.z];
    zBar.progress = ABS(acceleration.z);
}

@end
```

❶ The UIAccelerometer is a singleton object, so we grab a reference to the singleton rather than allocate and initialize a new instance of the class.

❷ We set the update interval to 0.1, hence the accelerometer:didAccelerate: method will be called 10 times every second.

❸ We declare that this class is the delegate for the UIAccelerometer.

❹ We implement the accelerometer:didAccelerate: delegate method and use it to set the X, Y, and Z labels to the raw accelerometer readings each time it is called. The progress bar values are set to the absolute value (the value without regard to sign) of the accelerometer reading.

OK, you're done. Before you click the Run button, make sure you've configured the project to deploy onto your iPhone or iPod touch to test it. Since this application makes use of the accelerometer, and iPhone Simulator doesn't have one, you're going to have to test it directly on the device.

If all goes well, you should see something that looks a lot like Figure 4-4.

Determining Device Orientation

Apple provide an easy way of determining the device orientation, a call to UIDevice will return the current orientation of the device:

```
    UIDevice *device = [UIDevice currentDevice];
    UIDeviceOrientation orientation = device.orientation;
```

This call will return a UIDeviceOrientation that can be: UIDeviceOrientationUnknown, UIDeviceOrientationPortrait, UIDeviceOrientationPortraitUpsideDown, UIDeviceOrientationLandscapeLeft, UIDeviceOrientationLandscapeRight or UIDeviceOrientationFaceUp. The sensor underlying this call is the accelerometer, and you'll see later in this chapter how to retrieve the device orientation directly from the raw accelerometer readings.

 As of the time of writing under iOS 4.3 the device does not correctly report a proper orientation when your application is first launched, with UIDevice returning null when queried.

Figure 4-4. The Accelerometer application running on an iPhone 4 sitting face-up on my desk, measuring a 1-gravity acceleration straight down

Lets go ahead and modify the Accelerometer application to display the device orientation. Click on the *AccelerometerViewController.h* interface file to open it in the Standard Editor and add the following code, highlighted below, to the class interface:

```
@interface AccelerometerViewController :
     UIViewController <UIAccelerometerDelegate> {

    IBOutlet UILabel *xLabel;
    IBOutlet UILabel *yLabel;
    IBOutlet UILabel *zLabel;

    IBOutlet UIProgressView *xBar;
    IBOutlet UIProgressView *yBar;
    IBOutlet UIProgressView *zBar;
    IBOutlet UILabel *orientationLabel;

    UIAccelerometer *accelerometer;
}

- (NSString *)stringFromOrientation:(UIDeviceOrientation) orientation;

@end
```

We're going to display the current orientation using in a `UILabel`, so we're going to have to write a convenience method `stringFromOrienation:` to convert the `UIDeviceOrienta tion` type returned by the `UIDevice` to an `NSString` to display in that label.

Make sure you've saved your changes, and click on the corresponding *Accelerometer-ViewController.m* implementation file and add the following method:

```
- (NSString *)stringFromOrientation:(UIDeviceOrientation) orientation {

    NSString *orientationString;
    switch (orientation) {
        case UIDeviceOrientationPortrait:
            orientationString = @"Portrait";
            break;
        case UIDeviceOrientationPortraitUpsideDown:
            orientationString = @"Portrait Upside Down";
            break;
        case UIDeviceOrientationLandscapeLeft:
            orientationString = @"Landscape Left";
            break;
        case UIDeviceOrientationLandscapeRight:
            orientationString = @"Landscape Right";
            break;
        case UIDeviceOrientationFaceUp:
            orientationString = @"Face Up";
            break;
        case UIDeviceOrientationFaceDown:
            orientationString = @"Face Down";
            break;
        case UIDeviceOrientationUnknown:
            orientationString = @"Unknown";
            break;
        default:
            orientationString = @"Not Known";
            break;
    }
    return orientationString;
}
```

Once you have added the `stringFromOrienation:` method, add the following code, highlighted below, to the existing `accelerometer:didAccelerate:` method in the same class:

```
- (void)accelerometer:(UIAccelerometer *)meter
      didAccelerate:(UIAcceleration *)acceleration {

    xLabel.text = [NSString stringWithFormat:@"%f", acceleration.x];
    xBar.progress = ABS(acceleration.x);

    yLabel.text = [NSString stringWithFormat:@"%f", acceleration.y];
    yBar.progress = ABS(acceleration.y);

    zLabel.text = [NSString stringWithFormat:@"%f", acceleration.z];
    zBar.progress = ABS(acceleration.z);
```

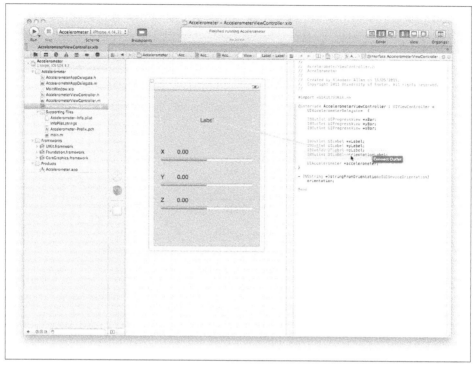

Figure 4-5. Connecting the orientation outlet to the UI

```
    UIDevice *device = [UIDevice currentDevice];
    orientationLabel.text = [self stringFromOrientation:device.orientation];
}
```

Make sure you've saved your changes, and click on the *AccelerometerViewController.xib* file to open it in Interface Builder. Drag and drop a UILabel from the Object Library into the View. Go ahead and resize and center up the text using the Attributes inspector from the Utilities panel.

Close the Utilities panel and open the Assistant Editor, which should show the corresponding interface file for the view controller. Control-click and drag and connect the UILabel element to the orientationLabel outlet in your code, as in Figure 4-5.

Save your changes, and click Run button in the Xcode toolbar to compile and deploy your application to your device. If all goes well you should see something much like Figure 4-6. As you move the device around, the label will update itself to reflect the current device orientation.

Determining Device Orientation Directly Using the Accelerometer

Instead of querying UIDevice you can use the raw accelerometer readings to determine the device orientation directly using the atan2 function as shown below:

Figure 4-6. The Accelerometer application reporting the device orientation

```
float x = -[acceleration x];
float y = [acceleration y];
float angle = atan2(y, x);
```

 For any real arguments x and y that are not both equal to zero, atan2(y, x) is the angle in radians between the positive x-axis of a plane and the point given by the specified coordinates on it. The angle is positive for counter-clockwise angles, and negative for clockwise angles.

Let's go ahead and modify the `accelerometer:didAccelerate:` method to calculate the orientation. Click on the *AccelerometerViewController.m* implementation file to open it in the Standard Editor and replace these lines:

```
UIDevice *device = [UIDevice currentDevice];
orientationLabel.text = [self stringFromOrientation:device.orientation];
```

with the code highlighted below:

```
- (void)accelerometer:(UIAccelerometer *)meter
    didAccelerate:(UIAcceleration *)acceleration {

    xLabel.text = [NSString stringWithFormat:@"%f", acceleration.x];
    xBar.progress = ABS(acceleration.x);
```

```
yLabel.text = [NSString stringWithFormat:@"%f", acceleration.y];
yBar.progress = ABS(acceleration.y);

zLabel.text = [NSString stringWithFormat:@"%f", acceleration.z];
zBar.progress = ABS(acceleration.z);

float x = -[acceleration x];
float y = [acceleration y];
float angle = atan2(y, x);

if(angle >= -2.25 && angle <= -0.75) {
    orientationLabel.text =
        [self stringFromOrientation:UIInterfaceOrientationPortrait];
} else if(angle >= -0.75 && angle <= 0.75){
    orientationLabel.text =
        [self stringFromOrientation:UIInterfaceOrientationLandscapeRight];
} else if(angle >= 0.75 && angle <= 2.25) {
    orientationLabel.text =
        [self
            stringFromOrientation:UIInterfaceOrientationPortraitUpsideDown];
} else if(angle <= -2.25 || angle >= 2.25) {
    orientationLabel.text =
        [self stringFromOrientation:UIInterfaceOrientationLandscapeLeft];
}

}
```

If you save your changes, and click on the Run button to rebuild and deploy your application onto your device, there should see little or no change in the application's operation. However, having access to each component of the orientation opens up many opportunities for creating tilt-based controls.

Obtaining Notifications when Device Orientation Changes

In addition to directly querying the UIDevice object for the current orientation, a program can request to be notified of changes in the device's orientation by registering itself as an observer.

We can once again modify the Accelerometer application to make use of this feature. Open the *AccelerometerViewController.m* file in the Standard Editor and delete the code added in the previous section from the accelerometer:didAccelerate: method.

If you quickly rebuild the application at this point and deploy it to your device you will see that the UILabel now reads "Label" and will no longer be updated as the device orientation changes.

Once you've confirmed that, add the following method:

```
-(void) viewWillAppear:(BOOL) animated{
    [super viewWillAppear:animated];
    [[UIDevice currentDevice] beginGeneratingDeviceOrientationNotifications];

    [[NSNotificationCenter defaultCenter]
```

```
    addObserver:self
      selector:@selector(receivedRotation:)
        name:UIDeviceOrientationDidChangeNotification
      object:nil];
}
```

Here we ask the `UIDevice` class to start generating orientation notifications, and we register the view controller as an observer. Next add the selector method, shown below:

```
-(void) receivedRotatation:(NSNotification*) notification {
    UIDevice *device = [UIDevice currentDevice];
    orientationLabel.text = [self stringFromOrientation:device.orientation];
}
```

Here we simply update the `UILabel` with the device orientation every time a `UIDevi ceOrientationDidChangeNotification` event is received.

Finally we need to remember to remove the program as an observer to stop the generation of messages during the tear down of our view controller. Add the following method to your code:

```
-(void) viewWillDisappear:(BOOL) animated{
    [super viewWillDisappear:animated];
    [[NSNotificationCenter defaultCenter] removeObserver: self];
    [[UIDevice currentDevice] endGeneratingDeviceOrientationNotifications];
}
```

If you once again save your changes and click on the Run button to rebuild and deploy the application to your device, you will again see little or no change in the application's operation.

Which Way Is Up?

A useful thing to know a lot of the time is the answer to the question "which way is up?" You can use the same method used earlier to determine the device orientation and graphically show this in the View.

First you're going to need an image of an arrow. Download, or draw in the graphics package of your choice, an image of an arrow pointing to the left on a transparent background. Save it as, or convert it to, a PNG format file. Drag-and-drop this into your Xcode Project remembering to tick the "Copy items into destination group's folder (if needed)" check box in the pop up dialog that appears when you drop the files into Xcode (see Figure 4-7).

Click on the *AccelerometerViewController.xib* file to open it, and drag-and-drop a `UII mageView` from the Object Library onto your View. Position it below the three `UIProg ressBar` elements, and resize the bounding box to be a square using the Size inspector of the Utility Pane. In the Attributes inspector of the Utility Pane, change the Image property to be the arrow image that you added to your project. Set the View mode to be "Aspect Fit". Uncheck the "Opaque" box in the Drawing section so that the arrow

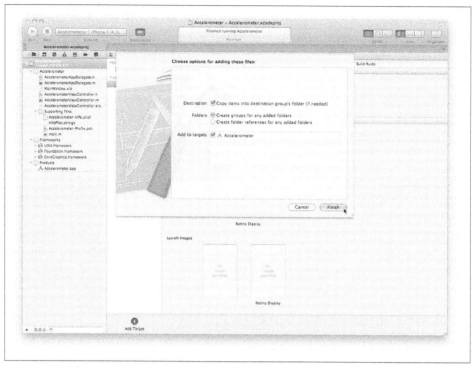

Figure 4-7. Adding an arrow image to the Accelerometer project

is rendered correctly with a transparent background. Finally, use the "Image" drop-down to select the arrow.png image to be displayed in the UIImageView (see Figure 4-8).

Close the Utility Pane and open the Assistant Editor. Control-click and drag from the UIImageView in your View to the **arrowImage** outlet in the Assistant Editor, as in Figure 4-9, and add an **arrowImage** outlet.

After doing so your interface file should look as below:

```
@interface AccelerometerViewController :
    UIViewController <UIAccelerometerDelegate> {

    IBOutlet UILabel *xLabel;
    IBOutlet UILabel *yLabel;
    IBOutlet UILabel *zLabel;

    IBOutlet UIProgressView *xBar;
    IBOutlet UIProgressView *yBar;
    IBOutlet UIProgressView *zBar;

    IBOutlet UIImageView *arrowImage;
    IBOutlet UILabel *orientationLabel;

    UIAccelerometer *accelerometer;
```

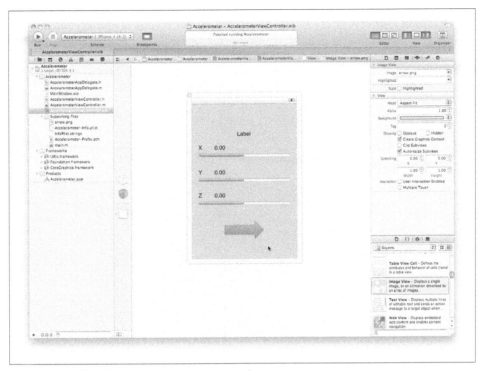

Figure 4-8. Adding the UIImageView to your interface

}

Close the Assistant Editor and switch to the Standard Editor. Go ahead and click on the *AccelerometerViewController.m* implementation file. Add the code highlighted below to the accelerometer:didAccelerate: method:

```
- (void)accelerometer:(UIAccelerometer *)meter
      didAccelerate:(UIAcceleration *)acceleration {

    xLabel.text = [NSString stringWithFormat:@"%f", acceleration.x];
    xBar.progress = ABS(acceleration.x);

    yLabel.text = [NSString stringWithFormat:@"%f", acceleration.y];
    yBar.progress = ABS(acceleration.y);

    zLabel.text = [NSString stringWithFormat:@"%f", acceleration.z];
    zBar.progress = ABS(acceleration.z);

    float x = -[acceleration x];
    float y = [acceleration y];
    float angle = atan2(y, x);
    [arrowImage setTransform:CGAffineTransformMakeRotation(angle)];
}
```

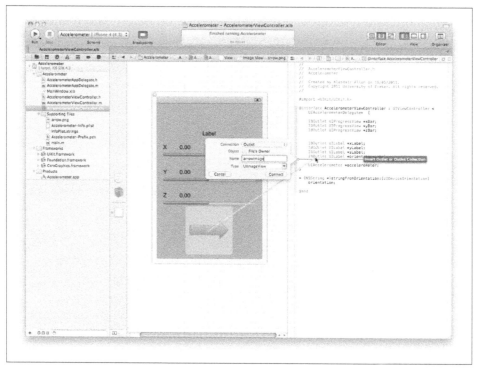

Figure 4-9. Adding a outlet to your code

That's it. Save your changes again and click on the Run button to compile and deploy the application to your device. Keep the device face towards you and rotate it in flat plane, you should see that the arrow moves as you do so, keeping its orientation pointing upwards (see Figure 4-9).

Convenience Methods for Orientation

Apple provides convenience methods to determine whether the current device orientation is portrait:

```
UIDevice *device = [UIDevice currentDevice];
UIDeviceOrientation orientation = device.orientation;
BOOL portrait = UIDeviceOrientationIsPortrait( orientation );
```

or landscape:

```
BOOL landscape = UIDeviceOrientationIsLandscape( orientation );
```

These methods return YES if the device is in portrait or landscape mode respectively; otherwise they return NO.

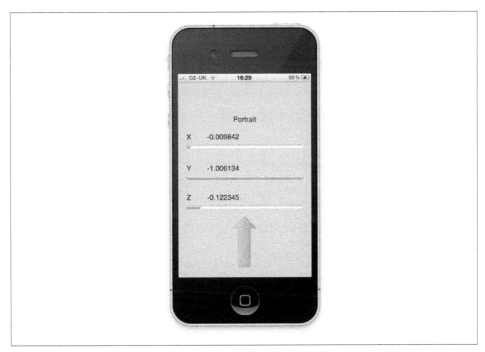

Figure 4-10. The arrow points upwards (device held vertically in front of the user)

Detecting Shaking

Apple's shake-detection algorithm analyses eight to ten successive pairs of raw accelerometer triplet values and determines the angle between these readings. If the change in angular velocity between successive data points is large then the algorithm determines that a `UIEventSubtypeMotionShake` has occurred, and the `motionBegan:withEvent:` delegate method is called. Conversely, if the change in angular velocity is small and a shake event has been triggered, the `motionEnded:withEvent:` delegate method is called.

> The iPhone is better at detecting side-to-side rather than front-to-back or up-and-down motions. Take this into account in the design of your application.

There are three motion delegate methods, mirroring the methods for gesture handling: `motionBegin:withEvent:`, `motionEnded:withEvent:` and `motionCancelled:withEvent:`. The first indicates the start of a motion event, the second the end of this event. You cannot generate a new motion event for a second (or two) following the first event. The final delegate method is called when a motion is interrupted by a system event, such as an incoming phone call.

Let's go ahead and add shake detection to our Accelerometer application. You'll need to add another UILabel to the UI that will change depending on the motion event status. Click on the *AccelerometerViewController.h* interface file to open it in the Standard Editor and add another UILabel marked as an IBOutlet to the class definition:

```
@interface AccelerometerViewController : UIViewController <UIAccelerometerDelegate> {

    IBOutlet UILabel *xLabel;
    IBOutlet UILabel *yLabel;
    IBOutlet UILabel *zLabel;

    IBOutlet UIProgressView *xBar;
    IBOutlet UIProgressView *yBar;
    IBOutlet UIProgressView *zBar;
    IBOutlet UILabel *orientationLabel;
    IBOutlet UILabel *shakeLabel;
    IBOutlet UIImageView *arrowImage;

    UIAccelerometer *accelerometer;

}
```

Save your changes and click on the *AccelerometerViewController.m* implementation file to open it in the Xcode editor.

The easiest way to ensure that the view controller receives motion events is to promote it to First Responder in the viewDidAppear: method. Remember to make the controller resign as first responder when the view goes away. Add the viewDidAppear: method and modify the existing viewWillDisappear: method as highlighted below. Use the canBe comeFirstResponder method to indicate that the view controller can indeed become the First Responder:

```
- (BOOL)canBecomeFirstResponder {
    return YES;
}

- (void)viewDidAppear:(BOOL)animated {
    [super viewDidAppear:animated];
    [self becomeFirstResponder];
}

-(void) viewWillDisappear: (BOOL) animated{
    [super viewWillDisappear:animated];
    [[NSNotificationCenter defaultCenter] removeObserver: self];
    [[UIDevice currentDevice] endGeneratingDeviceOrientationNotifications];
    [self resignFirstResponder];
}
```

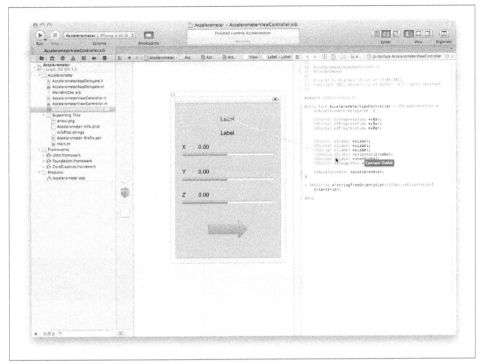

Figure 4-11. Connecting the shakeLabel outlet

Save your changes and click on the *AccelerometerViewController.xib* file for the last time. Drag-and-drop a `UILabel` into the View from the Object Library and connect it to the `shakeLabel` outlet as in Figure 4-11.

Save your changes and return to the *AccelerometerViewController.m* file in the editor, and add the following delegate methods to the implementation:

```
- (void)motionBegan:(UIEventSubtype)motion withEvent:(UIEvent *)event {
    if ( motion == UIEventSubtypeMotionShake ) {
        shakeLabel.text = @"SHAKE";
        shakeLabel.textColor = [UIColor redColor];
    }
    return;
}

- (void)motionEnded:(UIEventSubtype)motion withEvent:(UIEvent *)event {
    if ( motion == UIEventSubtypeMotionShake ) {
        shakeLabel.text = @"NO SHAKE";
        shakeLabel.textColor = [UIColor greenColor];
    }
    return;
}

- (void)motionCancelled:(UIEventSubtype)motion withEvent:(UIEvent *)event {
    if ( motion == UIEventSubtypeMotionShake ) {
```

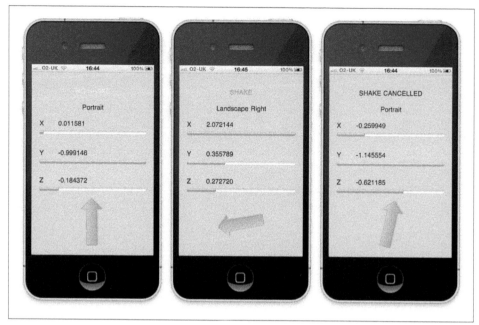

Figure 4-12. Shake detection on the iPhone

```
        shakeLabel.text = @"SHAKE CANCELLED";
        shakeLabel.textColor = [UIColor blackColor];
    }
    return;
}
```

Save your changes and click on the Run button in the Xcode toolbar. After the application is built and deployed to your device, try shaking the phone. You should see something very much like Figure 4-12.

CHAPTER 5

Using the Magnetometer

The magnetometer is a magnetoresistive permalloy sensor found in the iPhone 3GS, iPhone 4 and iPad 2, in addition to the accelerometer. The iPhone 3GS uses the AN-203 integrated circuit produced by Honeywell, while the iPhone 4 and iPad 2 make use of the newer AKM8975 produced by AKM Semiconductor. The sensor is located towards the top right hand corner of the device, and measures fields within a ±2 gauss (200 microtesla) range, and is sensitive to magnetic fields of less than 100 microgauss (0.01 microtesla).

 The Earth's magnetic field is roughly 0.6 gauss (60 microtesla). The field around a rare earth magnet can be 14,000 gauss or more.

The magnetometer measures the strength of the magnetic field surrounding the device. In the absence of any strong local fields, these measurements will be of the ambient magnetic field of the Earth, allowing the device to determine its "heading" with respect to the geomagnetic North Pole and act as a digital compass. The geomagnetic heading and true heading relative to the geographical North Pole can vary widely, by several tens of degrees depending on your location.

About the Magnetometer

Combining the heading (yaw) information (see Figure 5-1) returned by this device with the roll and pitch information returned by the accelerometer will let you determine the true orientation of the device in real time.

As well as reporting the current location, the CLLocationManager class can, in the case where the hardware supports it, report the current heading of the device. If location updates are also enabled, the location manager returns both true heading and magnetic heading values. If location updates are not enabled, the location manager returns only the magnetic heading value.

Figure 5-1. Using the magnetometer (a.k.a. the digital compass) in the iPhone 3GS you can determine the heading (yaw) of the device

 Magnetic heading updates are available even if the user has switched off location updates in the Settings application. Additionally, users are **not** prompted to give permission to use heading data, as it is assumed that magnetic heading information cannot compromise user privacy. On an enabled device the magnetic heading data should therefore always be available to your application.

As mentioned previously, the magnetometer readings will be affected by local magnetic fields, so the CLLocationManager may attempt to calibrate its heading readings by dis-

Figure 5-2. The Heading Calibration Panel

playing a heading calibration panel (see Figure 5-2) before it starts to issue update messages.

However, before it does so it will call the `locationManagerShouldDisplayHeadingCali bration:` delegate method:

```
- (BOOL)locationManagerShouldDisplayHeadingCalibration:
    (CLLocationManager *)manager {
    return YES;
}
```

If you return `YES` from this method, the `CLLocationManager` will pop up the calibration panel on top of the current window. The calibration panel prompts the user to move the device in a figure-eight pattern so that Core Location can distinguish between the Earth's magnetic field and any local magnetic fields. The panel will remain visible until calibration is complete or until you dismiss it by calling the `dismissHeadingCalibra tionDisplay:` method in the `CLLocationManager` class.

Writing a Magnetometer Application

Let's go ahead and implement a simple view-based application to illustrate how to use the magnetometer. Open Xcode and start a new iPhone project, select a View-based Application template, and name the project "Compass" when prompted for a filename.

Since you'll be making use of the Core Location framework, the first thing you need to do is add it to our new project. Click on the Compass project file in the Project navigator window on the right in Xcode, select the Target and click on the Build Phases tab, click on the Link with Libraries drop down and click on the + button to open the file pop-up window. Select *CoreLocation.framework* from the list of available frameworks and click the Add button.

You're going to build an application that will act as a compass, so you're going to need an image of an arrow to act as the compass needle. Download or draw in the graphics package of your choice, an image of an arrow pointing upwards on a transparent background. Save or convert it to, a PNG format file. Drag-and-drop this into the Xcode Project, remembering to tick the "Copy items into destination group's folder (if needed)" check box in the pop up dialog that appears when you drop the files into Xcode.

Click on *CompassViewController.xib* file to open it in Interface Builder. Drag and drop a UIImageView from the Object Library into the View, positioning it roughly in the center of your window, resizing the bounding box to be a square, as in Figure 5-3. In the Attributes inspector of the Utilities pane set the View mode to be "Aspect Fit", uncheck the "Opaque" checkbox in the Drawing section, and select the arrow image that you added to your project in the Image drop down.

Next drag-and-drop four UILabel elements from the Object Library into the View, position the four labels as in Figure 5-3, and change the text in the left most two to read "Magnetic Heading:" and "True Heading:".

Close the Utility pane and switch from the Standard to the Assistant Editor. Control-Click and drag from the two right most UILabel elements to the assistant editor to create a magneticHeadingLabel and trueHeadingLabel outlet, and then again for the UIImage View to create an arrowImage outlet, see Figure 5-3.

Then click on the *CompassViewController.h* interface file and go ahead and declare the class as a CLLocationManagerDelegate, remembering to import the *CoreLocation.h* header file. After doing so the interface should look like this:

```
#import <UIKit/UIKit.h>
#import <CoreLocation/CoreLocation.h>

@interface CompassViewController : UIViewController
    <CLLocationManagerDelegate> {

    IBOutlet UIImageView *arrowImage;
    IBOutlet UILabel *magneticHeadingLabel;
    IBOutlet UILabel *trueHeadingLabel;
}

@end
```

Save your changes, and click on the corresponding *CompassViewController.m* implementation file. Uncomment the viewDidLoad method and the following code to the

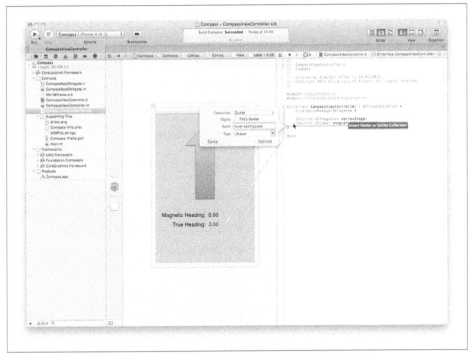

Figure 5-3. Connecting the outlets to the UI elements in Interface Builder

implementation. This will create an instance of the `CLLocationManager` class, and will send both location and heading update messages to the designated delegate class:

```
- (void)viewDidLoad {
    [super viewDidLoad];
    CLLocationManager *locationManager = [[CLLocationManager alloc] init];
    locationManager.delegate = self;
    if( [CLLocationManager locationServicesEnabled] &&
        [CLLocationManager headingAvailable]) {❶
        [locationManager startUpdatingLocation];
        [locationManager startUpdatingHeading];
    } else {
        NSLog(@"Can't report heading");
    }
}
```

❶ It is more important to check whether heading information is available than it is to check whether location services are available. The availability of heading information is restricted to the latest generation of devices.

You can (optionally) filter the heading update messages using an angular filter. Changes in heading of less than this amount will not generate an update message to the delegate, for example:

```
locationManager.headingFilter = 5;  // 5 degrees
```

The default value of this property is kCLHeadingFilterNone. You should use this value if you want to be notified of all heading updates. In this example, leave the filter set to the default value. However if you want to filter messages from Core Location this way, add the above line to your viewDidLoad method inside the if-block:

```
if( [CLLocationManager locationServicesEnabled] &&
    [CLLocationManager headingAvailable]) {
    [locationManager startUpdatingLocation];
    [locationManager startUpdatingHeading];
    locationManager.headingFilter = 5;  // 5 degrees
} else {
    ... code ...
}
```

The CLLocationManagerDelegate protocol calls the locationManager:didUpdateHeading: delegate method when the heading is updated. You're going to use this method to update the user interface. Add the following code to your view controller:

```
- (void)locationManager:(CLLocationManager*)manager
    didUpdateHeading:(CLHeading*)newHeading {

    if (newHeading.headingAccuracy > 0) {
        float magneticHeading = newHeading.magneticHeading;
        float trueHeading = newHeading.trueHeading;❶

        magneticHeadingLabel.text =
            [NSString stringWithFormat:@"%f", magneticHeading];
        trueHeadingLabel.text =
            [NSString stringWithFormat:@"%f", trueHeading];

        float heading = -1.0f * M_PI * newHeading.magneticHeading / 180.0f;
        arrowImage.transform = CGAffineTransformMakeRotation(heading);
    }
}
```

❶ If location updates are also enabled, the location manager returns both true heading and magnetic heading values. If location updates are not enabled, or the location of the device is not yet known, the location manager returns only the magnetic heading value and the value returned by this call will be –1.

Save your changes, then click on the Run button in the Xcode toolbar to deploy your new application to your device. If you hold the device in "Face Up" or "Portrait" mode you should see something very similar to Figure 5-4 below.

As it stands our application has a critical flaw. If the user orientates the device into Landscape Mode, the reported headings will be incorrect, or at least look incorrect to the user.

Determining the Heading in Landscape Mode

The magnetic and true headings are correct when the iPhone device is held like a traditional compass, in portrait mode, if the user rotates the device, the heading readings

---

Figure 5-4. The Compass application running on the iPhone 3GS

will still be in the original frame of reference. Even though the user has not changed the direction they are facing the heading values reported by the device will have changed. You're going to have to correct for orientation before reporting headings back to the user, see Figure 5-5.

In the Project navigator, click on the *CompassViewController.xib* file to open it in Interface Builder, then drag-and-drop another UILabel from the Object Library in the Utility pane into the View window. Use the Assistant Editor connect the label to a new outlet in the *CompassViewController.h* interface file, as in Figure 5-6.

After doing so, the interface file should look as below:

```
@interface CompassViewController :
    UIViewController <CLLocationManagerDelegate> {

    IBOutlet UILabel *trueHeadingLabel;
    IBOutlet UILabel *magneticHeadingLabel;
    IBOutlet UILabel *orientationLabel;
    IBOutlet UIImageView *arrowImage;

}
```

We're going to use this to report the current device orientation as we did in the Accelerometer application in Chapter 4.

Figure 5-5. The "real" heading of the user when they are holding the device in Landscape mode is the reported heading + 90 degrees

Close the Assistant Editor and reopen the *CompassViewController.h* interface file in the Standard Editor. Go ahead and add the following convenience methods to the class definition:

```
- (float)magneticHeading:(float)heading
     fromOrientation:(UIDeviceOrientation) orientation;
- (float)trueHeading:(float)heading
     fromOrientation:(UIDeviceOrientation) orientation;
- (NSString *)stringFromOrientation:(UIDeviceOrientation) orientation;
```

Save your changes, and open the *CompassViewController.m* implementation file. Since the CLHeading object is read only and you can't modify it directly, you'l need to add the following method to correct the magnetic heading for the device orientation:

```
- (float)magneticHeading:(float)heading
     fromOrientation:(UIDeviceOrientation) orientation {
```

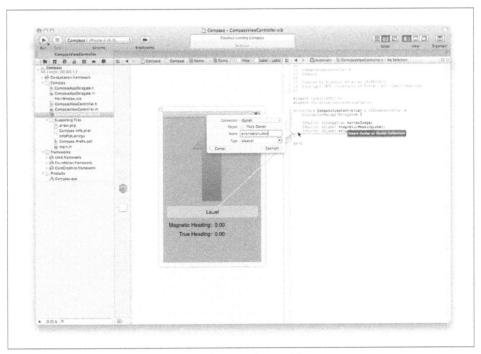

Figure 5-6. Connecting the orientation label in Interface Builder

```
float realHeading = heading;
switch (orientation) {❶
    case UIDeviceOrientationPortrait:
        break;
    case UIDeviceOrientationPortraitUpsideDown:
        realHeading = realHeading + 180.0f;
        break;
    case UIDeviceOrientationLandscapeLeft:
        realHeading = realHeading + 90.0f;
        break;
    case UIDeviceOrientationLandscapeRight:
        realHeading = realHeading - 90.0f;
        break;
    default:
        break;
}
while ( realHeading > 360.0f ) {
    realHeading = realHeading - 360;
}
return realHeading;
}
```

❶ The UIDeviceOrientationFaceUp and UIDeviceOrientationFaceDown orientation cases are undefined and the user is presumed to be holding the device in UIDeviceOrientationPortrait mode.

You will also need to add a corresponding method to correct the true heading:

```
- (float)trueHeading:(float)heading
      fromOrientation:(UIDeviceOrientation) orientation {

    float realHeading = heading;
    switch (orientation) {
        case UIDeviceOrientationPortrait:
            break;
        case UIDeviceOrientationPortraitUpsideDown:
            realHeading = realHeading + 180.0f;
            break;
        case UIDeviceOrientationLandscapeLeft:
            realHeading = realHeading + 90.0f;
            break;
        case UIDeviceOrientationLandscapeRight:
            realHeading = realHeading - 90.0f;
            break;
        default:
            break;
    }
    while ( realHeading > 360.0f ) {
        realHeading = realHeading - 360;
    }
    return realHeading;
}
```

Finally, add the `stringFromOrientation:` method from the previous section Chapter 5. We'll use this to update the `orientationLabel` outlet:

```
- (NSString *)stringFromOrientation:(UIDeviceOrientation) orientation {

    NSString *orientationString;
    switch (orientation) {
        case UIDeviceOrientationPortrait:
            orientationString = @"Portrait";
            break;
        case UIDeviceOrientationPortraitUpsideDown:
            orientationString = @"Portrait Upside Down";
            break;
        case UIDeviceOrientationLandscapeLeft:
            orientationString = @"Landscape Left";
            break;
        case UIDeviceOrientationLandscapeRight:
            orientationString = @"Landscape Right";
            break;
        case UIDeviceOrientationFaceUp:
            orientationString = @"Face Up";
            break;
        case UIDeviceOrientationFaceDown:
            orientationString = @"Face Down";
```

```
                break;
            case UIDeviceOrientationUnknown:
                orientationString = @"Unknown";
                break;
            default:
                orientationString = @"Not Known";
                break;
        }
        return orientationString;
    }
```

Return to the locationManager:didUpdateHeading: delegate method and modify the lines highlighted below to use the new methods and update the headings depending on the device orientation:

```
- (void)locationManager:(CLLocationManager*)manager
        didUpdateHeading:(CLHeading*)newHeading {

    UIDevice *device = [UIDevice currentDevice];
    orientationLabel.text = [self stringFromOrientation:device.orientation];

    if (newHeading.headingAccuracy > 0) {
        float magneticHeading = [self magneticHeading:newHeading.magneticHeading
                                    fromOrientation:device.orientation];
        float trueHeading = [self trueHeading:newHeading.trueHeading
                              fromOrientation:device.orientation];

        magneticHeadingLabel.text =
            [NSString stringWithFormat:@"%f", magneticHeading];
        trueHeadingLabel.text = [NSString stringWithFormat:@"%f", trueHeading];

        float heading = -1.0f * M_PI * newHeading.magneticHeading / 180.0f;❶
        arrowImage.transform = CGAffineTransformMakeRotation(heading);
    }
}
```

❶ You should still use the directly reported newHeading.magneticHeading for the compass needle rather than the adjusted heading. Otherwise the compass will not point correctly.

Make sure you've saved all the changes to the implementation file and click on the Run button in the Xcode toolbar to deploy the application onto the device. If all goes well you should see the same compass display as before. However this time if you rotate the display, sees Figure 5-7, the heading values should be the same irrespective of the device orientation.

Although I have not discussed or implemented it here, if the CLLocationManager object encounters an error, it will call the locationManager:didFailWithError: delegate method:

```
- (void)locationManager:(CLLocationManager *)manager
        didFailWithError:(NSError *)error {

    if ([error code] == kCLErrorDenied) {
```

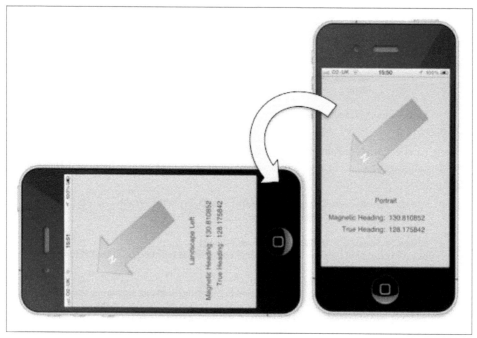

Figure 5-7. Heading values are now the same irrespective of orientation

```
        // User has denied the application's request to use location services.
        [manager stopUpdatingHeading];

    } else if ([error code] == kCLErrorHeadingFailure) {
        // Heading could not be determined
    }
}
```

Measuring a Magnetic Field

To use the device to measure a magnetic field—for instance that of a small bar magnet, or the field generated by an electric current in a wire—you should first make a zero-point measurement of the ambient magnetic field of the Earth. Further readings should subtract this zero-point measurement.

When measuring, move the magnet to the device rather than moving the device to the magnet. Moving the device will cause the magnetic field of the Earth across the measuring sensor to change, which will spoil the zero point calibration you took earlier. If the device must be moved, only small movements should be attempted.

You can retrieve the raw magnetic field measurements along the X, Y and Z-axes by querying the CLHeading object passed to the locationManager:didUpdateHeading: method:

```
- (void)locationManager:(CLLocationManager *)manager
      didUpdateHeading:(CLHeading *)heading {

    double x = heading.x;
    double y = heading.y;
    double z = heading.z;

    double magnitude = sqrt(x*x + y*y + z*z);

    ... code ...
}
```

The values returned are normalized into a ±128 range, measured in microtesla (μT), representing the offset from the magnetic field lines measured by the magnetometer.

 Apple provides sample code that displays the raw x, y, and z magnetometer values, a plotted history of those values, and a computed magnitude (size or strength) of the magnetic field. The code can be downloaded from the iPhone developer website at *http://developer.apple.com/library/ios/#samplecode/Teslameter*.

CHAPTER 6

Using Core Motion

The iPhone 4, latest generation iPod touch, and the iPad have a vibrational gyroscope in addition to an accelerometer and a magnetometer. The MicroElectroMechanical (MEMs) gyroscope inside the iPhone 4 and the 4th generation iPod touch is the AGD8 2032, nearly identical to an off-the-shelf STMicroelectronics L3G4200D device The iPad 2 uses an AGD8 2103 sensor, also from STMicroelectronics. These models operate by making use of a plate called a "proof mass" that oscillates when a drive signal is applied to capacitor plates. When the user rotates the phone, the proof mass is displaced in the X, Y and Z directions and an ASIC processor measures the capacitance change of the plates. The capacitance variation is used to detect the angular rate applied to the package.

An accelerometer provides measurement of forces in the X, Y and Z-axes but it cannot measure rotation. On the other hand, since a gyroscope is a rate of change device, you are able to measure the change in rotations around an axis. By using both sensors in combination you can measure the movement of the device in a six degrees-of-freedom inertial system, allowing you to use dead reckoning to find the physical location (and orientation of the device) relative to an initial starting position.

 All inertial systems have an inherent drift, so dead reckoning should not be regarded as being stable over the long term.

Core Motion

The arrival of iOS 4 brought with it the new Core Motion framework; this new framework allows your application to receive motion data from both the accelerometer and (on the latest generation of devices) the gyroscope.

There is no support for Core Motion in the iOS Simulator, therefore all testing of your Core Motion related code must be done on the device. The code in this chapter will only work on devices that have a gyroscope, see Chapter 1 for more information.

With the `CMMotionManager` class you can start receiving accelerometer, gyroscope, and combined device motion events at a regular interval, or you can poll them periodically:

```
CMMotionManager *motionManager = [[CMMotionManager alloc] init];
if (!motionManager.isDeviceMotionAvailable) {
    NSLog(@"Device supports motion capture.");
}
```

Remember to release the manager after you're done with it:

```
[motionManager release];
```

The `CMMotionManager` class offers both the raw accelerometer and gyroscope data separately as well a combined `CMDeviceMotion` object that encapsulates the processed device motion data from both the accelerometer and the gyroscope. With this combined motion measurement Core Motion provides highly accurate measurements of device attitude, the (unbiased) rotation rate of a device, the direction of gravity on a device, and the acceleration that the user is giving to a device.

The rotation rate reported by the `CMDeviceMotion` object is different than that reported directly by the gyroscope. Even if the device is sitting flat on the table the gyro will not read zero. It will read some non-zero value that differs from device to device and over time due to changes in things like device temperature. Core Motion actively tracks and removes this bias from the gyro data.

Pulling Motion Data

The `CMMotionManager` class offers two approaches to obtaining motion data. The simplest way is to pull the motion data. Your application will start an instance of the manager class and periodically ask for measurements of the combined device motion:

```
[motionManager startDeviceMotionUpdates];
CMDeviceMotion *motion = motionManager.deviceMotion;
```

Although if you are only interested in the raw gyroscope data (or accelerometer data) you can also ask for those directly:

```
CMGyroData *gyro = motionManager.gyroData;
CMAccelerometerData *accel = motionManager.accelerometerData;
```

This is the most efficient method of obtaining motion data. However, if there isn't a natural timer in your application—such as a periodic update of your main view—then you may need an additional timer to trigger your update requests. Remember to stop the updates and release the motion manager after you're done with them:

```
[motionManager stopDeviceMotionUpdates];
[motionManager release];
```

 Your application should create only a single instance of the CMMotion Manager class. Multiple instances of this class can affect the rate at which an application receives data from the accelerometer and gyroscope.

Pushing Motion Data

Instead of using this simple pull methodology, you can specify an update interval and implement a block of code for handling the motion data. The manager class can then be asked to deliver updates using the NSOperationsQueue, which allows the handler to push the measurements to the application. For example:

```
motionManager.deviceMotionUpdateInterval = 1.0/60.0;
[motionManager startDeviceMotionUpdatesToQueue: queue withHandler: handler];
```

or similarly for the individual accelerometer and gyroscope data:

```
[motionManager startAccelerometerUpdatesToQueue:queue withHandler: handler];
[motionManager startGyroUpdatesToQueue:queue withHandler:handler];
```

With this second methodology you'll get a continuous stream of motion data, but there is a large increased overhead associated with implementing it (see Table 6-1). Your application may not be able to keep up with the associated data rate especially if the device is in rapid motion.

Table 6-1. Example CPU usage for Core Motion push updates at 100 and 20Hz[a]

	At 100Hz		At 20Hz	
	Total	Application	Total	Application
DeviceMotion	65%	20%	65%	10%
Accelerometer	50%	15%	46%	5%
Accel + Gyro	51%	10%	50%	5%

[a] Figures for an application running on an iPhone 4 running iOS 4.0 (Reproduced with permission. Credit: Jeffrey Powers, Occipital)

Using Core Motion's combined CMDeviceMotion object, as opposed to accessing the raw CMAccelerometer or CMGyroData objects, consumes roughly 15% more total CPU regardless of the update rate. The good news is that is not because of the gyroscope itself; reading both the accelerometer and gyroscope directly is not noticeably slower than reading the accelerometer on its own.

Because of this associated CPU overheads push is really only recommended for data collection applications where the point of the application is to obtain the motion data itself. However if your application needs to be rapidly updated as to device motion you can do this easily:

```
CMMotionManager *motionManager = [[CMMotionManager alloc] init];
motionManager.deviceMotionUpdateInterval = 1.0/60.0;

if (motionManager.deviceMotionAvailable ) {
    queue = [[NSOperationQueue currentQueue] retain];
    [motionManager startDeviceMotionUpdatesToQueue:queue
                withHandler:^ (CMDeviceMotion *motionData, NSError *error) {

        CMAttitude *attitude = motionData.attitude;
        CMAcceleration gravity = motionData.gravity;
        CMAcceleration userAcceleration = motionData.userAcceleration;
        CMRotationRate rotate = motionData.rotationRate;
        // handle data here......
    }];
} else {
    [motionManager release];
}
```

If we were interested solely in the raw gyroscope data we could do the following:

```
CMMotionManager *motionManager = [[CMMotionManager alloc] init];
motionManager.gyroUpdateInterval = 1.0/60.0;

if (motionManager.gyroAvailable) {
    queue = [[NSOperationQueue currentQueue] retain];
    [motionManager startGyroUpdatesToQueue:queue
                withHandler: ^ (CMGyroData *gyroData, NSError *error) {

        CMRotationRate rotate = gyroData.rotationRate;
        NSLog(@"rotate x = %f, y = %f, z = %f", rotate.x, rotate.y, rotate.z);
        // handle rotation-rate data here......
    }];

} else {
    [motionManager release];
}
```

If we want both the raw and gyroscope and accelerometer readings outside of the
CMDeviceMotion object, we could modify the above code as highlighted:

```
CMMotionManager *motionManager = [[CMMotionManager alloc] init];
motionManager.gyroUpdateInterval = 1.0/60.0;
motionManager.accelerometerUpdateInterval = 1.0/60.0;

if (motionManager.gyroAvailable && motionManager.accelerometerAvailable) {
    queue = [[NSOperationQueue currentQueue] retain];
    [motionManager startGyroUpdatesToQueue:queue
                withHandler: ^ (CMGyroData *gyroData, NSError *error) {
        CMRotationRate rotate = gyroData.rotationRate;
        NSLog(@"rotate x = %f, y = %f, z = %f", rotate.x, rotate.y, rotate.z);
        // handle rotation-rate data here......
    }];
    [motionManager startAccelerometerUpdatesToQueue:queue
                        withHandler: ^ (CMAccelerometerData *accelData,
                                        NSError *error) {
        CMAcceleration accel = accelData.acceleration;
```

```
        NSLog( @"accel x = %f, y = %f, z = %f", accel.x, accel.y, accel.z);
        // handle acceleration data here......
    }];
} else {
    [motionManager release];
}
```

Accessing the Gyroscope

Let's go ahead and implement a simple view-based application to illustrate how to use the gyroscope on its own before looking again at Core Motion and CMDeviceMotion. Open Xcode and start a new View-based Application iPhone project and name it "Gyroscope" when prompted for a filename.

Since we'll be making use of the Core Motion framework, the first thing we need to do is add it to our new project. Click on the project file at the top of the Project navigator window on the right in Xcode, select the Target and click on the Build Phases tab, click on the Link with Libraries drop down and click on the + button to open the file pop-up window. Select *CoreMotion.framework* from the list of available frameworks and click the Add button.

Now go ahead and click on the *GyroscopeViewController.xib* file to open it in Interface Builder. As you did for the accelerometer back in Chapter 4, you're going to build a simple interface to report the raw gyroscope readings. Go ahead and drag and drop three UIProgressView from the Object Library into the View window, then add two UILabel elements for each progress bar: one to hold the X, Y, or Z label and the other to hold the rotation measurements. After you do that, the view should look something a lot like Figure 6-1.

Go ahead and close the Utilities panel and click to open the Assistant Editor. Then Control-Click and drag from the three UIProgressView elements, and the three UILabel elements that will hold the measured values, to the *GyroscopeViewController.h* header file which should be displayed in the Assistant Editor on the right-hand side of the interface (see Figure 6-2).

This will automatically create and declare three UILabel and three UIProgressView variables as an IBOutlet. Since they aren't going to be used outside the class, there isn't much point in declaring them as class properties, which you'd do with a Control-click and drag from the element to outside the curly brace. After doing this, the code should look like this:

```
#import <UIKit/UIKit.h>

@interface GyroscopeViewController : UIViewController {

    IBOutlet UIProgressView *xBar;
    IBOutlet UIProgressView *yBar;
    IBOutlet UIProgressView *zBar;
```

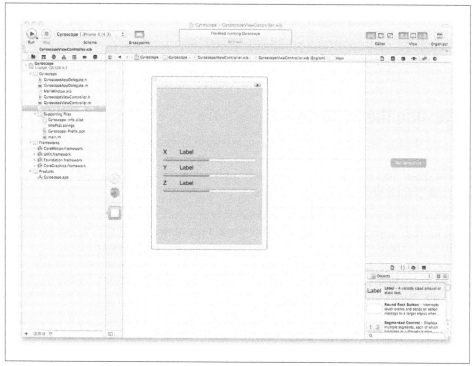

Figure 6-1. The Gyroscope UI

```
        IBOutlet UILabel *xLabel;
        IBOutlet UILabel *yLabel;
        IBOutlet UILabel *zLabel;
    }

    @end
```

Close the Assistant Editor, return to the Standard Editor and click on the *Gyroscope-ViewController.h* interface file. Go ahead and import the Core Motion header file, and declare a `CMMotionManager` and `NSOperationQueue` instance variables. Here's how the should look when you are done:

```
    #import <UIKit/UIKit.h>
    #import <CoreMotion/CoreMotion.h>

    @interface GyroscopeViewController : UIViewController {

        IBOutlet UIProgressView *xBar;
        IBOutlet UIProgressView *yBar;
        IBOutlet UIProgressView *zBar;

        IBOutlet UILabel *xLabel;
        IBOutlet UILabel *yLabel;
        IBOutlet UILabel *zLabel;
```

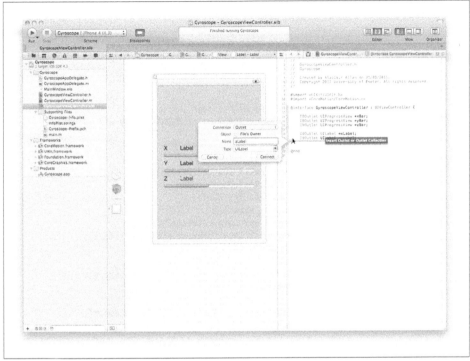

Figure 6-2. Connecting the UI elements to your code in Interface Builder

```
    CMMotionManager *motionManager;
    NSOperationQueue *queue;
}

@end
```

Make sure you've saved your changes and click on the corresponding *GyroscopeView-Controller.m* implementation file to open it in the Xcode editor. You don't actually have to do very much here, as Interface Builder handled most of the heavy lifting with respect to the UI, you just need to go ahead an implement the guts of the application to monitor the gyroscope updates:

```
@implementation GyroscopeViewController

- (void)dealloc {
    [xBar release];
    [yBar release];
    [zBar release];
    [xLabel release];
    [yLabel release];
    [zLabel release];
    [queue release];
    [super dealloc];
}
```

```objc
- (void)didReceiveMemoryWarning {
    [super didReceiveMemoryWarning];
}

#pragma mark - View lifecycle

- (void)viewDidLoad {
    [super viewDidLoad];
    motionManager = [[CMMotionManager alloc] init];
    motionManager.gyroUpdateInterval = 1.0/2.0; // Update every 1/2 second.

    if (motionManager.gyroAvailable) {
        NSLog(@"Gyroscope avaliable");
        queue = [[NSOperationQueue currentQueue] retain];
        [motionManager startGyroUpdatesToQueue:queue
            withHandler: ^ (CMGyroData *gyroData,
                            NSError *error) {
            CMRotationRate rotate = gyroData.rotationRate;
            xLabel.text = [NSString stringWithFormat:@"%f", rotate.x];
            xBar.progress = ABS(rotate.x);

            yLabel.text = [NSString stringWithFormat:@"%f", rotate.y];
            yBar.progress = ABS(rotate.y);

            zLabel.text = [NSString stringWithFormat:@"%f", rotate.z];
            zBar.progress = ABS(rotate.z);

        }];

    } else {
        NSLog(@"Gyroscope not available");
        [motionManager release];
    }
}

- (void)viewDidUnload {
    [motionManager stopGyroUpdates];
    [motionManager release];

    [xBar release];
    xBar = nil;
    [yBar release];
    yBar = nil;
    [zBar release];
    zBar = nil;
    [xLabel release];
    xLabel = nil;
    [yLabel release];
    yLabel = nil;
    [zLabel release];
    zLabel = nil;
    [super viewDidUnload];
}
```

```
- (BOOL)shouldAutorotateToInterfaceOrientation:
    (UIInterfaceOrientation)interfaceOrientation {

    return (interfaceOrientation == UIInterfaceOrientationPortrait);
}

@end
```

The CMRotationRate (*http://developer.apple.com/library/ios/documentation/CoreMo tion/Reference/CMAttitude_Class/Reference/Reference.html#//apple_ref/c/tdef/CMRo tationMatrix*) data structure provides the rate of rotations around X-, Y-, and Z-axes in units of radians per second. When inspecting the structure remember the right-hand rule to determine the direction of positive rotation. With your thumb in the positive direction on an axis, your fingers curl will give the positive rotation direction around that axis (see Figure 6-3). A negative rotation goes away from the tips of those fingers.

Let's test that: click on the Run button in the Xcode toolbar to build and deploy the application to your device (remember you can't test this code in the iOS Simulator). If all goes well you should see something much like Figure 6-4 as you roll the device around the Y-axis.

 As mentioned before the measurement of rotation rate encapsulated by a CMGyroData object is biased by various factors. You can obtain a much more accurate (unbiased) measurement by accessing the rotationRate (*http://developer.apple.com/library/ios/documentation/CoreMotion/Ref erence/CMDeviceMotion_Class/Reference/Reference.html#//apple_ref/ occ/instm/CMDeviceMotion/rotationRate*) property of CMDeviceMo tion (*http://developer.apple.com/library/ios/documentation/CoreMotion/ Reference/CMDeviceMotion_Class/Reference/Reference.html#//apple _ref/occ/cl/CMDeviceMotion*) if that is needed by your application.

Measuring Device Motion

Let's go ahead and build a similar application to the one above, but this time reporting the data exposed by the CMDeviceMotion object:

```
CMAttitude *attitude = motionData.attitude;
CMAcceleration gravity = motionData.gravity;
CMAcceleration userAcceleration = motionData.userAcceleration;
CMRotationRate rotate = motionData.rotationRate;
```

Open Xcode and start a new iPhone project, select a View-based Application template, and name the project "Motion" when prompted for a filename. As before, import the Core Motion framework into the project and then click on the *MotionViewControl-ler.xib* file to open it in Interface Builder, and then proceed to drag-and-drop UIProg ressBar and UILabel elements into your View in a similar manner as you did for the Gyroscope application earlier in the chapter. You'll need labels for the yaw, pitch and

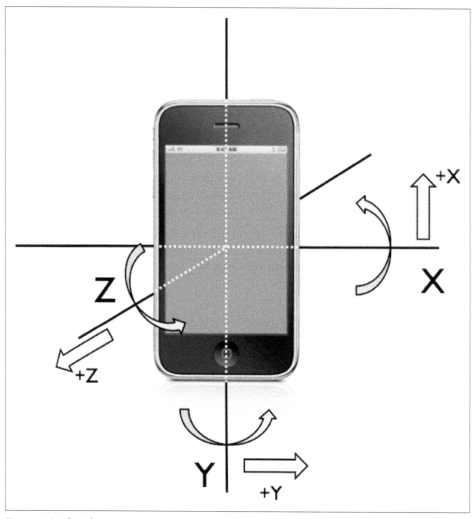

Figure 6-3. The iPhone gyroscope axes

roll values, along with progress bars and labels for the user acceleration, gravity and rotation values.

Once you've done this, go ahead and connect the various bars and labels as IBOutlet using the Assistant Editor into the *MotionViewController.h* interface file as in Figure 6-5.

Once you've done this, save your changes and open up the *MotionViewController.h* interface file in the Standard Editor. Go ahead and import the Core Motion framework:

```
#import <CoreMotion/CoreMotion.h>
```

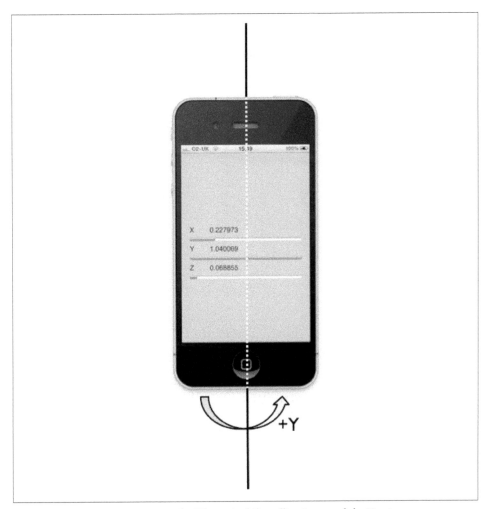

Figure 6-4. Measuring rotation on the iPhone 4 whilst rolling it around the Y-axis

For this example you're going to pull the device motion updates rather than push them using the NSOperationQueue and a handler block. Add the following instance variables to the view controller class:

```
CMMotionManager *motionManager;
NSTimer *timer;
```

Then in the corresponding *MotionViewController.m* implementation file, modify the viewDidLoad method as follows:

```
- (void)viewDidLoad {
    [super viewDidLoad];

    motionManager = [[CMMotionManager alloc] init];
```

Figure 6-5. The Motion application UI with the IBOutlet connected in Interface Builder to the MotionViewController.h interface file

```
motionManager.deviceMotionUpdateInterval = 1.0 / 10.0;
[motionManager startDeviceMotionUpdates];
if (motionManager.deviceMotionAvailable ) {
    timer = [NSTimer scheduledTimerWithTimeInterval:0.2f
                                    target:self
                                    selector:@selector(updateView:)
                                    userInfo:nil
                                    repeats:YES];
} else {
    [motionManager stopDeviceMotionUpdates];
    [motionManager release];
}
}
```

This will start the motion manager and begin polling for device motion updates. Add the following lines to the `viewDidUnload` method to corresponding stop the timer and updates:

```
[timer invalidate];
[motionManager stopDeviceMotionUpdates];
[motionManager release];
```

Once you have done this you should go ahead and implement the `updateView:` method that will be called by the `NSTimer` object:

```
-(void) updateView:(NSTimer *)timer {

    CMDeviceMotion *motionData = motionManager.deviceMotion;

    CMAttitude *attitude = motionData.attitude;
    CMAcceleration gravity = motionData.gravity;
    CMAcceleration userAcceleration = motionData.userAcceleration;
    CMRotationRate rotate = motionData.rotationRate;


    yawLabel.text = [NSString stringWithFormat:@"%2.2f", attitude.yaw];
    pitchLabel.text = [NSString stringWithFormat:@"%2.2f", attitude.pitch];
    rollLabel.text = [NSString stringWithFormat:@"%2.2f", attitude.roll];

    accelIndicatorX.progress = ABS(userAcceleration.x);
    accelIndicatorY.progress = ABS(userAcceleration.y);
    accelIndicatorZ.progress = ABS(userAcceleration.z);
    accelLabelX.text = [NSString stringWithFormat:@"%2.2f",userAcceleration.x];
    accelLabelY.text = [NSString stringWithFormat:@"%2.2f",userAcceleration.y];
    accelLabelZ.text = [NSString stringWithFormat:@"%2.2f",userAcceleration.z];

    gravityIndicatorX.progress = ABS(gravity.x);
    gravityIndicatorY.progress = ABS(gravity.y);
    gravityIndicatorZ.progress = ABS(gravity.z);
    gravityLabelX.text = [NSString stringWithFormat:@"%2.2f",gravity.x];
    gravityLabelY.text = [NSString stringWithFormat:@"%2.2f",gravity.y];
    gravityLabelZ.text = [NSString stringWithFormat:@"%2.2f",gravity.z];

    rotIndicatorX.progress = ABS(rotate.x);
    rotIndicatorY.progress = ABS(rotate.y);
    rotIndicatorZ.progress = ABS(rotate.z);
    rotLabelX.text = [NSString stringWithFormat:@"%2.2f",rotate.x];
    rotLabelY.text = [NSString stringWithFormat:@"%2.2f",rotate.y];
    rotLabelZ.text = [NSString stringWithFormat:@"%2.2f",rotate.z];

}
```

Save your changes and hit the Run button in the Xcode toolbar to build and deploy the application to your device. If all goes well you should see something much like Figure 6-6.

Comparing Device Motion with the Accelerometer

At this stage we can illustrate the difference between gravity and user-contributed acceleration values reported by Core Motion to the raw acceleration values reported by the UIAccelerometer, discussed back in Chapter 4.

Re-open the *MotionViewController.xib* file in Interface Builder and add another section to the UI, which will report the raw readings from the UIAccelerometer object. Go ahead and connect these three bars to IBOutlet instance variables in the *MotionViewController.h* interface file using the Assistant Editor as before, see Figure 6-7.

Figure 6-6. The Motion application running on an iPhone 4 sitting flat on the desk, with gravity in the negative Z-direction without any rotation or user acceleration

As you can see from Figure 6-7, I've changed the UIProgressView style from "Default" to "Bar" using the Attributes inspector in the Utility pane. This will help differentiate this section—data reported from the UIAccelerometer—from the other sections whose values are reported by the CMMotionManager.

Once that is done, close the Assistant Editor and open the *MotionViewController.h* interface file using the Standard Editor. Go ahead and declare the view controller as a UIAccelerometerDelegate and add a UIAccelerometer instance variable as shown here:

```
@interface MotionViewController : UIViewController <UIAccelerometerDelegate> {

    ...

    UIAccelerometer *accelerometer;
}

@end
```

before opening the corresponding implementation file. In the viewDidLoad method, add the following code to initialize the UIAccelerometer object. You should see Chapter 4 for more details on the UIAccelerometer class and associated methods:

```
- (void)viewDidLoad {
    [super viewDidLoad];
```

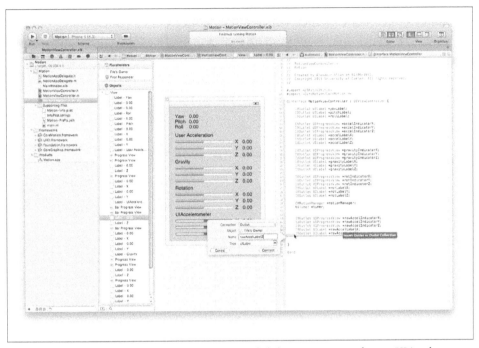

Figure 6-7. The additional UIProgressView and UILabel elements to report the raw UIAccelerometer readings to the user

```
motionManager = [[CMMotionManager alloc] init];
motionManager.deviceMotionUpdateInterval =  1.0 / 10.0;
[motionManager startDeviceMotionUpdates];
if (motionManager.deviceMotionAvailable ) {
    timer = [NSTimer scheduledTimerWithTimeInterval:0.2f
                                   target:self
                                 selector:@selector(updateView:)
                                 userInfo:nil
                                  repeats:YES];
} else {
    [motionManager stopDeviceMotionUpdates];
    [motionManager release];
}

accelerometer = [UIAccelerometer sharedAccelerometer];
accelerometer.updateInterval = 0.2f;
accelerometer.delegate = self;
}
```

After doing this all you need to do is add the accelerometer:didAccelerate: delegate method:

```
- (void)accelerometer:(UIAccelerometer *)meter
    didAccelerate:(UIAcceleration *)acceleration {
```

Figure 6-8. The Motion application running on an iPhone 4 sitting flat on the desk

```
rawAccelLabelX.text =
  [NSString stringWithFormat:@"%2.2f", acceleration.x];
rawAccelIndicatorX.progress = ABS(acceleration.x);

rawAccelLabelY.text =
  [NSString stringWithFormat:@"%2.2f", acceleration.y];
rawAccelIndicatorY.progress = ABS(acceleration.y);

rawAccelLabelZ.text =
  [NSString stringWithFormat:@"%2.2f", acceleration.z];
rawAccelIndicatorZ.progress = ABS(acceleration.z);
}
```

Click on the Run button in the Xcode toolbar to build and deploy the application to the device as before. If all goes well you should see something much like Figure 6-8.

Move the device around and you can see how the raw accelerometer values and the derived gravity and user acceleration values correspond to each other.

CHAPTER 7

Going Further

I've covered a fair amount of ground in the last few chapters, and you should now have a solid grasp of the basics of handling the sensor data produced by the hardware.

The iPhone SDK

Predictably in a book talking about sensors I've focused on the parts of the SDK that will be most helpful, and allow you to use the basic sensor hardware in your own applications. But even there I've left out a lot in an attempt to simplify and get you started quickly, especially when it comes to audio. A more in-depth look at the iPhone SDK is available in *Programming iOS 4* (*http://oreilly.com/catalog/9781449388430*), by Matt Neuburg (O'Reilly).

Geolocation and Maps

The iPhone is one of the most popular devices for geolocation: users use it for everything from driving directions to finding a restaurant close to them. As a developer, you can get in on the geolocation game by using the Core Location framework, one of the most powerful and interesting frameworks in the iPhone SDK. It abstracts the details of determining a user's location, and does all the heavy lifting for you behind the scenes. From there you can use the MapKit framework to embed maps directly into your views, and then go ahead and annotate those maps. I'll deep-dive into both these topics in upcoming title *Geolocation in iOS* (*http://oreilly.com/catalog/9781449308445*), by Alasdair Allan (O'Reilly).

Third-Party SDKs

The same book will investigate third-party geo-SDKs such as the Skyhook Wireless Local Faves (*http://www.skyhookwireless.com/localfaves*) and Spot Rank (*http://www .skyhookwireless.com/spotrank*) SDKs, along with coverage of SimpleGeo (*https://sim plegeo.com*) and SG Context and Places.

Speech Recognition

I covered basic manipulation of the audio hardware, but moving on from this you might be thinking about integrating speech recognition into your application. At least until Apple gets round to adding this to the official iOS SDK, the best way to do this is probably using the CMU Pocketsphinx and CMU Flite libraries. There are actually two fairly good Objective-C wrappers to the libraries, these are VocalKit (*https://github.com/KingOfBrian/VocalKit*) and OpenEars (*http://www.politepix.com/openears*). Of the two, at least at the time of writing, OpenEars probably has is the best documentation which may be a deciding factor if you're not a expert in speech recognition.

Computer Vision

The Open Source Computer Vision (OpenCV) Library (*http://opencv.willowgarage .com/wiki*) is a collection of routines intended for real-time computer vision, released under the BSD License, free for both private and commercial use. The library has a number of different possible applications including object recognition and tracking. We delve into computer vision and face recognition in the upcoming title *Augmented Reality in iOS* (*http://oreilly.com/catalog/9781449308506*), by Alasdair Allan (O'Reilly).

While you wait you might want to take a look at some of the sample code from this title which is already on the web at *http://programmingiphonesensors.com/pages/oscon .html*.

Augmented Reality

Unsurprisingly perhaps, the same title will also take a close look at Augmented Reality, which has become one of the killer applications for the iOS platform. The book walks you through building a simple location-aware AR toolkit, and some of the sample code is already online at *http://programmingiphonesensors.com/masterclass/theclass.html*.

If you're interested in AR you might also want to take a look at the associated video masterclass on iOS Sensors which features me, amongst other things, walking you through the AR toolkit code (*http://oreilly.com/catalog/0636920014508*).

 The video masterclass *Making use of iPhone and iPad Location Sensors* was filmed using Xcode 3 and iOS 4.0. While the code is still fine, the step-by-step walkthroughs in Xcode are somewhat out of date.

External Accessories

While the iOS platform comes with a growing range of sensors; GPS, accelerometers, magnetometers and most recently gyroscopes. They also have a (near-)ubiquitous data connection, whether via a local wireless hotspot or via carrier data, and user positioning via multiple methods including GPS. The device makes an excellent hub for a distributed sensor network.

However until recently it was actually quite difficult to interface these otherwise interesting devices into a standard serial interface, as the iPhone's proprietary dock connector is a major stumbling block.

All this has changed. In the upcoming title *iOS and Sensor Networks* (*http://oreilly.com/ catalog/9781449308483*) by Alasdair Allan (O'Reilly) we'll discuss using the MFi approved Redpark Serial Cable. This is an officially Apple approved route, and makes use of Apple's own External Accessory Framework to connect your device to any standard serial (RS-232) capable device. In addition to this we will go on to discuss other methods to use the phone as the hub of a sensor network, and part of the Internet of Things.

About the Author

Alasdair Allan is the author of *Learning iPhone Programming* (*http://learningiphone programming.com*), *Programming iPhone Sensors* (*http://programmingiphonesensors .com*), *Basic Sensors in iOS* (*http://oreilly.com/catalog/9781449308469*), *Geolocation iOS* (*http://oreilly.com/catalog/9781449308445*), *iOS Sensor Apps with Arduino* (*http:// oreilly.com/catalog/9781449308483*), and *Augmented Reality in iOS* (*http://oreilly.com/ catalog/9781449308506*), all published by O'Reilly Media. He is a senior research fellow (*http://www.astro.ex.ac.uk/people/aa*) in Astronomy (*http://www.astro.ex.ac.uk*) at the University of Exeter (*http://www.exeter.ac.uk*). As part of his work there, he is building a distributed peer-to-peer network of telescopes (*http://www.estar.org.uk*) which, acting autonomously, will reactively schedule observations of time-critical events. Notable successes include contributing to the detection of the most distant object yet discovered, a gamma-ray burster at a redshift of 8.2 (*http://arxiv.org/abs/0906 .1577*). Alasdair also runs a small technology consulting business (*http://www.babilim .co.uk*) writing bespoke software, building open hardware, and providing training. He writes for O'Reilly Radar (*http://radar.oreilly.com/aallan/index.html*), and sporadically writes in his own blog, The Daily ACK (*http://www.dailyack.com*), about things that interest him, or more frequently provides commentary in 140 characters or less on Twitter (*http://twitter.com/aallan*).

Colophon

The animal on the cover of *Basic Sensors in iOS* is a Malay fox-bat.

The cover image is from Lydekker's *Royal Natural History*. The cover font is Adobe ITC Garamond. The text font is Linotype Birka; the heading font is Adobe Myriad Condensed; and the code font is LucasFont's TheSansMonoCondensed.

Get even more for your money.

Join the O'Reilly Community, and register the O'Reilly books you own. It's free, and you'll get:

- $4.99 ebook upgrade offer
- 40% upgrade offer on O'Reilly print books
- Membership discounts on books and events
- Free lifetime updates to ebooks and videos
- Multiple ebook formats, DRM FREE
- Participation in the O'Reilly community
- Newsletters
- Account management
- 100% Satisfaction Guarantee

Signing up is easy:

1. **Go to: oreilly.com/go/register**
2. **Create an O'Reilly login.**
3. **Provide your address.**
4. **Register your books.**

Note: English-language books only

To order books online:
oreilly.com/store

For questions about products or an order:
orders@oreilly.com

To sign up to get topic-specific email announcements and/or news about upcoming books, conferences, special offers, and new technologies:
elists@oreilly.com

For technical questions about book content:
booktech@oreilly.com

To submit new book proposals to our editors:
proposals@oreilly.com

O'Reilly books are available in multiple DRM-free ebook formats. For more information:
oreilly.com/ebooks

Spreading the knowledge of innovators | oreilly.com

The information you need, when and where you need it.

With Safari Books Online, you can:

Access the contents of thousands of technology and business books

- Quickly search over 7000 books and certification guides
- Download whole books or chapters in PDF format, at no extra cost, to print or read on the go
- Copy and paste code
- Save up to 35% on O'Reilly print books
- **New!** Access mobile-friendly books directly from cell phones and mobile devices

Stay up-to-date on emerging topics before the books are published

- Get on-demand access to evolving manuscripts.
- Interact directly with authors of upcoming books

Explore thousands of hours of video on technology and design topics

- Learn from expert video tutorials
- Watch and replay recorded conference sessions

O'REILLY®